Computational Fluid Dynamics and the Theory of Fluidization

Huilin Lu · Dimitri Gidaspow · Shuyan Wang

# Computational Fluid Dynamics and the Theory of Fluidization

## Applications of the Kinetic Theory of Granular Flow

Springer

Huilin Lu
School of Science and Engineering
Harbin Institute of Technology
Harbin, Heilongjiang, China

Shuyan Wang
School of Petroleum Engineering
Northeast Petroleum University
Daqing, Heilongjiang, China

Dimitri Gidaspow
Department of Chemical and Biological Engineering
Illinois Institute of Technology
Chicago, IL, USA

ISBN 978-981-16-1557-3 ISBN 978-981-16-1558-0 (eBook)
https://doi.org/10.1007/978-981-16-1558-0



This Springer imprint is published by the registered company Springer Nature Singapore Pte Ltd.
The registered company address is: 152 Beach Road, #21-01/04 Gateway East, Singapore 189721, Singapore

# Preface

The widespread application of fluidization in industry demands an increase in efficiency and the development of predictive numerical simulations of industrial processes for design and scale-up purposes. In spite of today's advances in computational capability, considerable challenges remain due to the complexity of fluidized bed processes. This book is intended to address some of these issues and provide fundamental knowledge and needed design and scale-up tools for graduate students in different branches of science and engineering (i.e., chemical, biological, mechanical, and environmental engineers, etc.) and scientists and engineers already working in areas related to fluid-particle fluidized beds.

This book provides an up-to-date description of the formulation, implementation, and applications of Eulerian-Eulerian two-fluid model (TFM) modeling with kinetic theory of granular flow (KTGF) in fluidized beds. It is an integrated text that deals with theoretical and practical concepts of KTGF-based TFM and its industrial applications. KTGF-based TFM has found a wide range of applications in nearly all systems dealing with solids in various fields of science, engineering and technology such as chemical, food, pharmaceutical, biochemical, mechanical, energy, material, and mineral engineering. However, the prime concern of this book is to provide a more comprehensive treatment of KTGF-based TFM in chemical and process engineering with applications in combustion, chemical reactors, and multiphase flow systems.

The kinetic theory of granular flow, which is based on the oscillation of the particles, uses a granular temperature equation to determine the turbulent kinetic energy of the particles, assumes a distribution function for instantaneous particle velocity, and defines a constitutive equation based on particle collision, interaction and fluctuation. In fact, the kinetic theory approach for granular flow allows the determination of, for example, particle-phase stress, pressure, and viscosity in place of the empirical equations. After this developmental stage of granular theory, there have been several modifications to the constitutive equations such as extension of the kinetic theory for multi-type particles. This book provides a comprehensive approach to address the above-mentioned challenges and issues, and presents fundamental knowledge for design and scale-up tools of gas-solid fluidized beds.

Chapter 1 deals with the derivation of the basic governing equations for conservation of mass, momentum, and energy for multiphase systems.

Chapter 2 provides continuity and momentum conservation equations and constitutive and boundary conditions for fluid-particle flow systems. The conservation equations and constitutive relations are general and can be applied to all regimes of fluid-particle flow, from a very dilute particle volume fraction to the packed bed regime. The fundamentals of the kinetic theory approach for derivation of constitutive equations for the regimes when particle collision is dominant and the frictional behavior of particles based on soil mechanics principles for dense particle flow are discussed. In addition, the kinetic theory approach has been extended to multi-type particulate flows.

Chapter 3 deals with the effects of the presence of particle clusters on modeling and simulation in circulating fluidized beds. This chapter introduces two approaches for considering the presence of clusters and large structures that have gained significant attention in the literature: the filtering or subgrid models and the heterogeneous structure-based drag models. The fundamental derivation of dynamic cluster structure-dependent drag model is presented to predict drag coefficient of heterogeneous structures. The predicted heterogeneous structure parameters are compared to experimental measurements in circulating fluidized beds.

Chapter 4 provides experimental verification of kinetic theory of granular flow in a vertical pipe. Particle velocity distributions were measured using a video-digital camera technique, and granular temperatures were calculated according to standard deviation of fluctuating velocity. Particle pressures were measured using a differential transducer. The solids volume fractions were measured using an X-ray and gamma-ray densitometers. The radial distribution functions were measured using the digital camera. Experiments showed that a relation exists among pressure, granular temperature, and density, analogous to the ideal gas law. There was an agreement between the kinetic-theory measurement and previous macroscopic viscosity measurements.

Chapter 5 introduces the basic concepts for the solution of nonlinear algebraic equations. The nonlinear integral equations summarize these concepts using the theory of functional analysis. The discretization methods for the convection-diffusion terms of conservation equations are described to derive the discrete transport equation, and the set of discretized equations is outlined. The under relaxation procedure used for stabilizing the calculations, the linear equation solvers and the calculation of residuals used for judging the convergence of iterations are described in the various steps of the solution algorithm.

Chapter 6 presents three case studies. Case 1 is CFD modeling and simulation of a circulating fluidized bed coal combustor and hybrid pulverized-fluidized bed coal combustor. Case 2 is CFD modeling and simulation of a diameter transformed fluidized bed chemical-looping combustion, circulating fluidized bed chemical-looping combustion and fluidized bed chemical looping reforming. Case 3 is numerical simulations of fluidized bed $CO_2$ capture.

The authors greatly appreciate several individuals who contributed in different ways to the completion of this book. In particular, the authors would like to acknowledge the contributions of the following: Dr. Liu Guodong and Wang Shuai of Harbin Institute of Technology for modeling of fluidized beds using kinetic theory of granular flow, Dr. Wang Lin for the modeling of multi-type particle flow using the kinetic theory approach, Dr. Bing Sun of UOP (Des Plaines, IL), Dr. Yuanxiang Wu, Dr. Reza Mostofi of UOP (Des Plaines, IL) and Dr. Diana Matonis for the tutorial and manual of the IIT CFD code, and Dr. Mitra Bahary for experimental measurements of fluidized beds. Thanks should be extended to Dr. Jiang Xiaoxue, Cai Wenjian, and Li Dan (Harbin Institute of Technology) for the development of the dynamic cluster structure-dependent drag model from modeling to applications.

We thank the National Science Foundation of China (NSFC) for their long-term financial support of this work, including No. 51776059 and 51876032, and National Key Research and Development Program No. 2018YFF0216002. The authors would like to thank and greatly appreciate Mrs. Zhang Nan and Springer for language editing this book.

Harbin, China — Huilin Lu
Chicago, USA — Dimitri Gidaspow
Daqing, China — Shuyan Wang

# Contents

# Chapter 1
# Introduction to Fluidization Basic Equations

## Nomenclature

| | |
|---|---|
| $D_{iss}$ | Energy dissipation by means of friction, kJ/s |
| $F$ | Body force, N |
| $h$ | Enthalpy per unit mass, J/kg |
| $\boldsymbol{I}$ | Identity matrix |
| $m$ | Mass, kg |
| $m_i'$ | Rate of production of phase i, kg/(s·m$^3$) |
| $P$ | Pressure, N/m$^2$ |
| $\bar{p}$ | Interaction forces between phases or momentum supply, N |
| $Q$ | Heat transfer, J |
| $S$ | Entropy, J/K |
| $T$ | Stress tensor |
| $t$ | Time, s |
| $V$ | System volume, m$^3$ |
| $v$ | Mean velocity vector, m/s |
| $U$ | Internal energy per unit mass, J/kg |
| $Y_{ik}$ | Weight fraction of component $k$ in phase $i$ |

## Greek Symbols

| | |
|---|---|
| $\alpha_i$ | Volume fraction of phase $i$ |
| $\beta$ | Interphase friction coefficient |
| $\mu$ | Viscosity, kg/m·s |
| $\lambda$ | Bulk viscosity, kg/m·s |
| $\rho$ | Density, kg/m$^3$ |
| $\Im$ | General property per unit volume general property per unit volume |



H. Lu et al., *Computational Fluid Dynamics and the Theory of Fluidization*,
https://doi.org/10.1007/978-981-16-1558-0_1

## 1.1 Introduction

The multiphase modeling framework represents a trade-off between accuracy and computational efforts for fluidized bed reactors. The average multi-fluid models are a direct multi-dimensional extension of the conventional fluidized bed reactor models with the addition of the momentum equations. The multi-fluid model is actually the only approach that can be used analyzing chemical reactive systems on industrial scales considering both inherent physics and feasible computational efforts. A fundamental task for chemical reaction engineers is the formulation of transport equations of gas phase and solids phase in the fluidized bed reactors. These models have been used determining various quantitative parameters of reactor performance such as velocities, temperature, production rate, conversion and selectivity. The ability to establish such quantitative links between simulations of fluidized bed reactor performance and input and operating variables is essential in optimizing the operating conditions in design and scale-up of fluidized bed reactors.

The Eulerian-Eulerian two-fluid models, where a relatively large number of particles are considered determining a continuous phase in the control volume formulating the governing model equations, denote the mass and momentum conservation equations as formulated in an Eulerian frame of reference for both the dispersed and continuous phases describing the time dependent motion. Conservation laws for mass, momentum, and energy for fluidized beds can be derived using the Reynolds transport theorem, as illustrated by Aris (1989). An advantage of the Eulerian-Eulerian two-fluid model is that in principle it can be used to compute any multiphase flow regime, provided that an adequate closure relation for the interfacial coupling terms are provided. Besides, when this model is applied to describe dense dispersed flows, the Eulerian-Eulerian approach is normally less computationally demanding in numerical simulations of fluidized bed reactors. As the flow in most operating chemical reactors is dense, the two-fluid models have been found to represent a trade-off between accuracy and computational efforts for practical applications.

For multiphase fluidized bed flow systems, the concept of the volume fraction for phase i, $\alpha_i$, appears. For a single-phase system, $\alpha_i = 1$, and these equations reduce themselves to those found in standard transport phenomena books, such as Bird et al. (2007). We need two concepts: conservation laws and constitutive equations. We show in this section how these conservation laws are obtained from the Reynolds transport theorem, in reference to Fig. 1.1.

The point $(x^0, y^0, z^0)$ represents the spatial coordinates of the mass at some fixed time $t^0$. Then, the spatial coordinates of the mass at any time, $t$, are given by:

$$x = x\left(t, x^o, y^o, z^o\right) \quad y = y\left(t, x^o, y^o, z^o\right) \quad z = z\left(t, x^o, y^o, z^o\right) \tag{1.1}$$

A property per unit volume $\Im(t, x)$ is defined, where $t$ is time and $x$ is the position vector such that

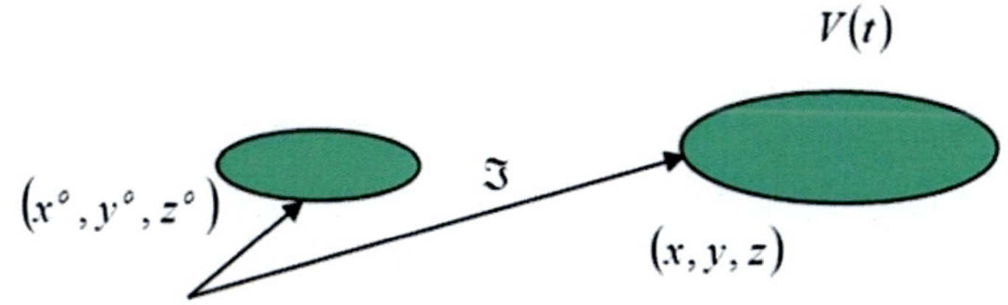


**Fig. 1.1** Motion for a system of constant mass

$$F(t) = \iiint\limits_{V(t)} \Im(t, x) dV. \tag{1.2}$$

*F(t)* is the system variable quantity that can change with time. The balance made on *F(t)* gives the Reynolds transport theorem

$$\frac{d}{dt} \iiint\limits_{V(t)} \Im(t, x) dV = \iiint\limits_{V(t)} \left( \frac{\partial \Im}{\partial t} + \nabla \cdot \Im v_i \right) dV \tag{1.3}$$

In multiphase flow, the volume occupied by phase $i$ cannot be occupied by any other phase at the same position in space at the same time. This distinction introduces the concept of the volume fraction of phase $i$, $\alpha_i$. The volume of phase *i*, $V_i$, in a system of volume *V*, is, as illustrated by Gidaspow (1994).

$$V_i = \iiint\limits_{V(t)} \alpha_i dV \quad \text{where} \quad \sum_{i=1}^{n} \alpha_i = 1 \tag{1.4}$$

Equation (1.3) defines into existence the volume fraction, $\alpha_i$, for phase *i*. Other authors have taken an alternative means of defining this quantity related to volume averaging. The volume fraction for the dispersed phase in some volume $\delta V$ is defined as $\alpha_d = \lim\limits_{\delta V \to \delta V^o} \delta V_d / \delta V$ where $\delta V_d$ is the volume of the dispersed phase in the volume and $\delta V^o$ is the limiting volume which ensures a stationary average, that is the volume should be large enough to contain a representative number of dispersed phase elements even if it is decreased or increased slightly.

## 1.2 Conservations Laws for Fluidization

The Reynolds transport theorem will be used to derive the multiphase balances of mass, momentum, and energy (Brennen 2005; Arastoopour et al. 2017).

### 1.2.1 Mass Balances

The mass of phase *i* in a volume element V moving with phase *i* velocity is as follows:

$$m_i = \iiint\limits_{V(t)} \alpha_i \rho_i dV \tag{1.5}$$

where, $\alpha_i$ is the volume fraction of phase *i* and $\rho_i$ is defined as the density of phase *i*. This volume fraction is the only new variable in multiphase flow. Traditionally, this volume fraction, $\alpha_i$, was derived by volume averaging as discussed above and by Jackson (2000). For multicomponent systems $\alpha_i$ is one, because it is assumed that molecules of component *i* occupy the same space at the same time. Such an approximation cannot be made in multiphase flow. In multiphase flow the volume fraction for phase *i* cannot be occupied by the remaining phases at the same position in space at the same time.

The balance on mass $m_i$ moving with the velocity $v_i$ is

$$\frac{dm_i}{dt^i} = \frac{d}{dt} \iiint \alpha_i \rho_i dV = \iiint m_i' dV \tag{1.6}$$

where $m_i'$ is the rate of production of phase *i*; $d/dt^i$ is the substantial derivative moving with phase *i* velocity.

The Reynolds transport theorem gives the continuity equation for phase *i*:

$$\frac{\partial(\alpha_i \rho_i)}{\partial t} + \nabla \cdot (\alpha_i \rho_i v_i) = m_i' \tag{1.7}$$

Conservation of mass requires that

$$\sum_{i=1}^{n} m_i' = 0 \tag{1.8}$$

### 1.2.2 Momentum Balances

The rate of change of momentum of phase *i* moving with velocity $v_i$ equals the forces acting on the system and can be mathematically expressed by the following:

$$\frac{d}{dt} \iiint\limits_{V(t)} \rho_i v_i \alpha_i dV = f_i \tag{1.9}$$

where, the force $f_i$ is given by

$$f_i = \oiint_{S(t)} T_i da + \iiint_{V(t)} \rho_i F_i \alpha_i dV + \overline{p}_i dV + m_i v_i dV \tag{1.10}$$

The first term represents the surfaces forces acting on the differential area *da*, the second term the external forces, the third term the interaction forces between phases, and the last term the force due to the phase change.

The stress tensor $T_i$ for phase *i* is composed of 9 components and can be written as follows.

$$T_i = \begin{pmatrix} T_{ixx} & T_{ixy} & T_{ixz} \\ T_{iyx} & T_{iyy} & T_{iyz} \\ T_{izx} & T_{izy} & T_{izz} \end{pmatrix} \tag{1.11}$$

Figure 1.2 represents the stress tensor acting on surfaces of volume of phase *i*. The Reynolds transport theorem is applied to the left side of Eq. (1.9) and the divergence theorem $\oiint_a T_i da = \iiint_V \nabla \cdot T_i dV)$ to the first term of Eq. (1.10). The result is the momentum balance for phase *i*:

$$\frac{\partial}{\partial t}(\alpha_i \rho_i) + \nabla \cdot (\alpha_i \rho_i v_i v_i) = \nabla \cdot T_i + \rho_i \alpha_i \mathbf{F}_i + \overline{p}_i + m_i' \mathbf{v}_i \tag{1.12}$$

Using the continuity equation Eq. (1.7), the momentum balance for phase *i* moving with phase velocity $v_i$ becomes

$$\rho_i \alpha_i \frac{dv_i}{dt^i} = \nabla \cdot T_i + \rho_i \alpha_i F_i + \overline{p}_i \tag{1.13}$$

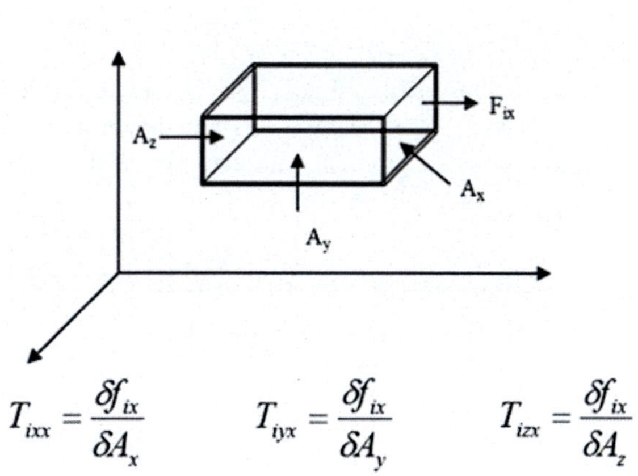


**Fig. 1.2** Stress tensor acting on surfaces of volume of phase *i*

where $\rho_i \alpha_i \frac{dv_i}{dt^i}$ is the acceleration of phase $i$, $\nabla \cdot T_i$ is the gradient of stress due to the surface forces, $\rho_i \alpha_i F_i$ is the body force, and $\overline{p_i}$ is the interaction forces between phases, called the momentum supply. Note that$\sum_{i=1}^{nphase} \overline{p}_i = 0$.

The interaction terms between phases given by the expression

$$\overline{p}_i = \sum_j \beta_j (v_j - v_k) \tag{1.14}$$

account for the drag force between the phases in terms of friction coefficients, $\beta_j$. Note that moving with the velocity $v_i$ there is no force due to the phase change, because the mass in the balance equation Eq. (1.9) moves with velocity $v_i$.

The simplest expression for the stress, $T_i$, is given by

$$T_i = -P_i I \tag{1.15}$$

Each phase $i$ has its own pressure $P_i$ similar to Euler's equation for single-phase flow.

#### 1.2.2.1 Incompressible Viscous Flow

To meet the requirement of objectivity for each phase $k$

$$T_k = T_k\left(\nabla^s v_k\right) \tag{1.16}$$

where, $\nabla^s v_k$ is the symmetrical gradient of $v_k$.

Linearization of $T_k$ gives

$$T_k = A_k I + B_k\left(\nabla^s v_k\right) \tag{1.17}$$

For incompressible fluids, $A_k$ is chosen to be the negative of the pressure of fluid phase $k$, $A_{k} = \_P_k = -k$, $A_k = \_P_k$ and the derivative of the traction, $T_k$, with respect to the symmetric gradient with viscosity of fluid $k$, is given by

$$B_k = 2\mu_k = \frac{\partial T_k}{\partial \nabla^s v_k} \tag{1.18}$$

The traction for phase $k$ is given by

$$T_k = -P_k I + 2\mu_k \nabla^s v_k \tag{1.19}$$

where

$$\nabla^s v_k = \frac{1}{2}\left[\nabla v_k + (\nabla v_k)^T\right] \tag{1.20}$$

#### 1.2.2.2 Incompressible Navier–Stokes Equation

Substitution of stress tensors for each phase $k$ into the momentum balance Eq. (1.12) gives the incompressible Navier–Stokes equation for each phase $k$:

$$\frac{\partial}{\partial t}(\alpha_k \rho_k \boldsymbol{v}_k) + \nabla \cdot (\alpha_k \rho_k \boldsymbol{v}_k \boldsymbol{v}_k) \\ - \nabla p_k - \nabla \overline{p}_k + \mu_k \nabla^2 v_k + \sum\nolimits_j \beta_j (v_j - v_k) + \rho_k \alpha_k \boldsymbol{F}_k + m'_k \boldsymbol{v}_k \tag{1.21}$$

The friction coefficients $\beta_j$ can be expressed in terms of standard drag coefficients (Gidaspow 1994; Syamlal et al. 2017). In this model the viscosity of each phase $k$, $\mu_k$, is an input into the model.

#### 1.2.2.3 Compressible Viscous Flow

For compressible viscous flow there is another parameter, the bulk viscosity of phase $k$, $\lambda_k$:

$$A_k = -P_k + \lambda_k \nabla \cdot v_k \tag{1.22}$$

Then the stress for phase $k$ is expressed as follows

$$T_k = -P_k I + 2\mu_k \nabla^s v_k + \lambda_k I \nabla \cdot v_k \tag{1.23}$$

where, $2\mu_k + 3\lambda_k = 0$, similar to that for single phase flow.

Hence the compressible Navier–Stokes Equation for each phase $k$ in substantial derivative form becomes as follows:

$$\alpha_k \rho_k \frac{dv_k}{dt^k} = \alpha_k \rho_k F_k + \sum_j \beta_j (v_j - v_k) - \nabla P_k + \nabla\left(2\mu_k \nabla^s v_k + \frac{2}{3}\mu_k I \nabla \cdot v_k\right) \tag{1.24}$$

### 1.2.3 Energy Balances

The energy balance for an open system with phase change can be written for phase $i$ as follows (Lyczkowski et al. 1982; Gidaspow and Jiradilok 2009).

$$\frac{dU_i}{dt^i} = \frac{dQ_i}{dt} - p_i\frac{dV_i}{dt^i} + D_{iss} + (U_i^n + \frac{p_i^n}{\rho_i^n})\frac{dm_i}{dt^i} \tag{1.25}$$

where, $U_i$ is the internal energy per unit mass; $Q_i$ is the heat per unit mass flowing into the system and the only work done by the system is the mechanical work due to the volume changes, $P_i dV_i$. With phase change there is inflow due to phase change, consisting of the energy inflow, $U_i{}^n$, and pressure, $P_i{}^n$. The energy dissipation by the system is $D_{iss} = \iiint_{V(t)} u_i dV$.

The rate of heat transfer, $Q_i$, is related to the rate of heat transfer by conduction, $q_i$, through the following relations:

$$-\frac{dQ_i}{dt} = \oiint_{A(t)} q_i\alpha_i da = \iiint_{V(t)} (\nabla \cdot \alpha_i q_i) dV \tag{1.26}$$

In Eq. (1.24) the rate of heat transfer, $q_i$, need not be multiplied by the volume fraction of phase $i$.

The Reynolds transport theorem applied to the energy balance Eq. (1.25) produces the energy equation for phase $i$ in the form

$$\frac{\partial(\alpha_i\rho_i U_i)}{\partial t} + \nabla \cdot (\alpha_i\rho_i v_i U_i) = -\nabla \cdot \alpha_i q_i - P_i\frac{\partial \alpha_i}{\partial t} - P_i\nabla \cdot \alpha_i v_i + h_i^n m_i' + u_i' \tag{1.27}$$

where $u_i^{'}$ is the internal energy per unit volume, $h_i{}^n$ is the enthalpy flowing into the system and $q_i$ is the rate of heat transfer by conduction.

There appears a strange term in Eq. (1.27), $P_i\frac{\partial \alpha_i}{\partial t}$, the work of expansion of volume fraction of phase $i$, not found in single phase flow.

#### 1.2.3.1 Enthalpy Representation

The energy balance expressed in terms of enthalpy of phase $i$ per unit of mass, $h_i$, in substantial derivative form becomes as follows:

$$\alpha_i\rho_i\frac{dh_i}{dt^i} = -\nabla \cdot \alpha_i q_i + \alpha_i\frac{dp_i}{dt^i} + U_i + m_i'(h_i^n - h_i) \tag{1.28}$$

Equation (1.27) reduces itself to that found in standard transport phenomena text books for $\varepsilon_i$ equal to 1 and $h_i{}^n = h_i$. The energy equation Eq. (1.28) in enthalpy form has no strange work term found in Eq. (1.25).

#### 1.2.3.2 Entropy Representation

The entropy balance for phase $i$ moving with the velocity $v_i$ is as follows

$$\frac{d}{dt}\iiint\limits_{V(t)} \alpha_i \rho_i S_i dV + \oiint\limits_{A(t)} \frac{q_i \alpha_i da}{T_i} - \iiint m_i' S_i dV = \iiint \sigma_i dV \tag{1.29}$$

where: $\frac{d}{dt}\iiint\limits_{V(t)} \alpha_i \rho_i S_i dV$ is the accumulation of entropy in the system, $\oint\limits_{A(t)} \frac{q_i \alpha_i da}{T_i}$ is the rate of entropy outflow due to flow of heat across area $A$, $\iiint m_i' S_i dV$ is the rate of entropy at equilibrium due to the phase change, and $\iiint \sigma_i dV$ is the rate of entropy production.

The combination of entropy balance Eq. (1.29) with the energy balance Eq. (1.25) generates the following equation for entropy production in the system, which is zero for reversible processes and positive for irreversible processes:

$$\sigma_i = -\frac{1}{T_i}\nabla \cdot \alpha_i q_i + \nabla \cdot \frac{q_i \alpha_i}{T_i} + \frac{U_i'}{T_i} = -\frac{q_i \alpha_i \cdot \nabla T_i}{T_i^2} + \frac{U_i'}{T_i} \geq 0 \tag{1.30}$$

### 1.2.4 Conservation of Species

To be able to treat a fluid in which a chemical reaction occurs, such as fluidized bed reactors, one must write mass balances for the individual components in a phase. The conservation of gas species gives the balance (Gidaspow 1994)

$$\frac{\partial(\alpha_i \rho_i Y_{ik})}{\partial t} + \nabla \cdot (\alpha_i \rho_i v_{ik} Y_{ik}) = m_{sik}' \tag{1.31}$$

$i = 1, 2, \ldots$n phases, and $k = 1, 2, \ldots$m components.

where $Y_{ik}$ is the weight fraction of component $k$ in phase $i$. The right-hand side of Eq. (1.29) has been written in a more natural way in terms of a specific mass transfer rate $\dot{m}_{si}$. The term $\dot{m}_{sik}$ includes the transfer of component $k$ from another phase and its creation by chemical reaction in phase $i$. The density $\rho_i$ is a function of pressure and enthalpy of phase $i$ and its composition.

A set of averaged partial differential equations describing a fluid-particles two-phase flow needs extra equations to close the system. The closure laws are of three types: constitutive, transfer and topological laws. The constitutive laws relate physical properties using axioms and experimental data. The transfer laws are empirical equations that describe the interaction between the phases occurring at the interface. The topological laws are needed to describe the spatial distribution of a flow variable. The different types of closure laws are simply referred to as constitutive laws. The key to accurate modeling lies to a large extent in the closure laws used. The considerable number of models has been developed to allow a coarser mesh and a longer time step to be used in numerical simulations, but, it may still be difficult to find suitable closure laws for flow of fluid-particles mixture. For the particulate phase, there are two quite different ways of modeling such flow parameters as dynamic viscosity and pressure of particles. One is to make empirical models based on the particle properties and the local gas volume fractions. These traditional models are relatively simple, and are thus easy to implement in a computer code. The other way to model these properties is to use the so-called kinetic theory of granular flow. Generally, these kinds of models are more complex and time-consuming to solve, but they are applicable for a wider range of problems. Nevertheless, the closure models become necessary for performance optimization, scale-up and design of industrial scale reactors.

## References

Arastoopour, H., Gidaspow, D., & Abassi, E. (2017). *Computational transport phenomena of fluid-particle systems*. Cham Switzerland: Springer.

Aris, R. (1989). *Vectors, tensors, and the Basic Equations of Fluid Mechanics*. New York: Dover Publications Inc.

Bird, R. B., Stewart, W. E., & Lightfoot, E. N. (2007). Transport phenomena. In: John Wiley and Sons (eds) 2nd edn. New York

Brennen, C. E. (2005). *Fundamentals of multiphase flows*. Cambridge: University Press.

Gidaspow, D. (1994). *Multiphase flow and fluidization*. Continuum and kinetic theory description: Academic Press Inc., San Diego California.

Gidaspow, D., & Jiradilok, V. (2009). *Computational techniques: The multiphase CFD approach to fluidization and green energy technologies*. New York: Nova Science Publishers.

Jackson, R. (2000). *The dynamics of fluidized particles*. Cambridge: University Press.

Lyczkowski, R. W., Gidaspow, D., & Solbrig, C. W. (1982). Multiphase flow models for nuclear, fossil and biomass energy production. Advances in Transport Processes, Vol. II, p 198–351, Majumdar AS and Mashelkar RA (Eds) Wiley Eastern Ltd New Delhi.

Syamlal, M., Musser, J., & Dietiker, J. F. (2017). Two-fluid model in MFIX, Multiphase Flow Handbook. In: Michaelides EE, Crowe CT, Schwarzkopf JD (eds) 2nd edn. Boca Raton, Florida, pp. 242–274.

# Chapter 2
# Constitutive Equations with Kinetic Theory of Granular Flow

## Nomenclature

| | |
|---|---|
| $A$ | Avogadro's number |
| $\mathbf{c}$ | Velocity space, m/s |
| $\vec{c}_i$ | Instantaneous particle velocity in *i* direction, m/s |
| $C$ | Fluctuation velocity, m/s |
| $d_{ip} = (d_i + d_p)/2$ | Average diameter of particles i and p, m |
| $d_p$ | Diameter of particle, m |
| $E$ | Internal energy, J |
| $e$ | Restitution coefficient |
| $e_{sw}$ | Particle–wall restitution coefficient |
| $f$ | Frequency distribution of velocities of particles |
| $f_s^0$ | Maxwellian velocity distribution function |
| $f_s^1$ | Dingle velocity distribution function |
| $f_{ip}^2$ | Pair velocity distribution function |
| $g$ | Gravity acceleration, $m/s^2$ |
| $g0$ | Radial distribution function |
| $g_{ip}$ | Radial distribution function at contact between particles i and p |
| $g_{ss}$ | Radial distribution function at contact between particles of the same phase |
| $\overline{I}$ | Identity tensor |
| $k$ | Unit vector connecting the centers of the two particles |
| $K_f$ | Turbulent kinetic energy of fluid, $m^2/s^2$ |
| $m_s$ | Mass of particle phase s, kg |
| $m_0 = m_i + m_p$ | Total mass of two colliding particles, kg |
| $n_s$ | Number density of phase s |
| $N_{ip}$ | Energy dissipation |
| $p_c$ | Critical state pressure, $N/m^2$ |
| $p_f$ | Frictional pressure, $N/m^2$ |



H. Lu et al., *Computational Fluid Dynamics and the Theory of Fluidization*,
https://doi.org/10.1007/978-981-16-1558-0_2

| | |
|---|---|
| $q_c$ | Collisional flux, J/s.m$^2$ |
| $q_k$ | Kinetic flux, J/s.m$^2$ |
| $Re_p$ | Particle Reynolds number |
| $\langle S \rangle$ | Average strain rate |
| $\boldsymbol{r}$ | Position |
| $S$ | Deviatoric stress tensor |
| $T$ | Average of random kinetic energy |
| $\mathbf{v}$ | Hydrodynamic velocity, m/s |
| $v_s$ | Mean velocity of phase s, m/s |
| $v_{ts}$ | Solid particle terminal velocity, m/s |

### Subscripts

| | |
|---|---|
| $col$ | Collisional |
| $cr$ | Critical |
| $fr$ | Frictional |
| $g$ | Gas phase |
| $kin$ | Kinetic |
| $max$ | Maximum |
| $s$ | Solid phase |
| $T$ | Total |
| w | Wall |

### Greek Symbols

| | |
|---|---|
| $\chi_{ip}$ | Collisional flux, J/s·m$^2$ |
| $\gamma_{ip}$ | Collisional source of granular temperature, J/m$^2$·s |
| $\mu$ | Viscosity, kg/m·s |
| $\mu_{ss}$ | Granular viscosity of phase s, kg/m·s |
| $\mu_{ip}$ | Mixture granular viscosity, kg/m·s |
| $\rho_s$ | Solid density of phase s, kg/m$^3$ |
| $\theta_s$ | Fluctuating granular energy, m$^2$/s$^2$ |
| $\theta_m$ | Mixture fluctuating granular energy, m$^2$/s$^2$ |
| $\tau_s$ | Solid phase shear stress, N/m$^2$ |
| $\psi_s$ | Property of particle |
| $\theta$ | Granular temperature (random kinetic energy per unit mass), m$^2$/s$^2$ |
| $\phi$ | Specularity coefficient |
| $\phi_{gs}$ | Granular energy exchange between phases |
| $\alpha_s$ | Volume fraction of solids |
| $\rho_s$ | Solids density, kg/m$^3$ |
| $II_{2D}$ | Second invariant of the deviatoric stress tensor |
| $\gamma$ | Shear rate, 1/s |
| $\beta$ | Interface momentum exchange coefficient |
| $\gamma$ | Collisional energy dissipation, J/m$^3$·s |
| $\xi$ | Solid phase bulk viscosity, kg/m·s |

## 2.1 Introduction

A granular material is defined as a collection of a large number of solid particles. Granular flow might thus be referred to as a powder flow. In a granular material there are primarily three mechanisms by which the bulk stresses are generated. These are (i) dry friction, (ii) transport of momentum by particle translation like in dilute gases, and (iii) momentum transport by particle interactions like in dense gases and liquids. It was usually assumed that although all three mechanisms co-exist in most flow regimes, one of them will normally play a predominant role. If the motion occurs slowly, particles will stay in contact and interact frictionally with their neighbors over long periods of time. This particle motion is characterized as the quasi-static regime of granular flow. The dry friction stresses are of the quasi-static, rate independent, Coulomb, type as described in the soil mechanics. To model such flows the momentum equation is normally extended including an extra stress so that the total stress is the linear sum of frictional, kinetic and collisional contributions, each calculated independently from constitutive expressions derived for the limits of pure frictional, pure kinetic, and pure collisional interactions, respectively. It is noted that the quasi-static stress models are still of empirical nature. Due to these additional stresses, the boundary conditions might also be adapted to the particular application. On the other hand, the other extreme is the rapid-flow regime in which the motion becomes rapid enough to transfer enough energy to particles. In this way the rapid flow regime might expand until every particle in the granular mass is moving independently. These flows are of great interest to chemical reactor engineers because the fluidized bed reactors are operated within this regime.

To model the flows rapid-flow regime an analogy between particle collisions in suspensions and molecular collisions in dense gases using kinetic theory was proposed by Bagnold (1954). By employing a simple expression for the collision frequency of particles, Bagnold derived a relation for the repulsive pressure of particles for uniform shear flow. Nevertheless, the importance of the velocity fluctuations and the means by which they could be included in a properly formulated continuum theory were exploited (Ogawa et al. 1980). The velocity of each particle is decomposed into a sum of the mean velocity of the bulk material and an apparently random component to describe the motion of the particle relative to the mean. Furthermore, a fluctuation particle temperature, commonly referred to as the granular temperature, was defined as the mean kinetic energy of the particle velocity fluctuations. Moreover, the mechanical energy of a granular flow is first transformed into random particle motion and then dissipated into internal energy.

Originally, the kinetic theory was developed by Chapman and Cowling (1961) for gases to predict the behavior of molecules whose interaction energies are conserved. For a granular flow, Savage and Jeffrey (1981) were the first to suggest that the kinetic theory for dense gases as proposed by Enskog (1922) might be useful determining solids stress tensor. They considered a collection of smooth, rigid, elastic, spherical particles in Couette flow. From elementary kinetic theory they recognize that in order to deduce explicit expressions for the solids stress tensor, it is necessary to determine the form of the complete pair-distribution function at collision. It was assumed that the pair-distribution function can be expressed as a product of a configurational pair-correlation function and two single particle distribution functions. The configurational pair-correlation function represents a generalization of the more familiar radial distribution function. The particle distribution function is restricted to a gas at equilibrium in which there is no mean deformation so spatial homogeneity of the mean flow prevails. Furthermore, the single distribution function of each particle was approximated by the Maxwellian. Jenkins and Savage (1983) extended the preliminary kinetic theory considering slightly inelastic particles. They used Maxwell's equations of transfer to derive local expressions for the balance of granular mass, granular linear momentum, granular temperature, and integral expressions for the stress, energy flux and energy dissipation terms. In order to evaluate these integrals and thereby close the transport equations it is again necessary to determine the complete pair-distribution function. In this case the preliminary distribution function proposed by Savage and Jeffrey (1981) for elastic particles was modified reformulating the configurational pair-correlation function expression so that it depends linearly upon the rate of deformation. Besides, the granular temperature equation was taken into account to determine the local rate of change of the fluctuating energy considering the energy dissipation rate due to inelastic particle collisions. In addition, the complete pair-distribution function is approximated by a symmetric average obtained shifting the pair of spatial points at which it is evaluated through a first order Taylor expansion. A balance law was formulated for the granular temperature that related the rate of change to the production by the mean flow shear, the heat like flux from one point to another in the flow, and the dissipation into thermal energy. The energy transport in granular flows can thus be interpreted as follows. The source of all the energy is the work done on the system, either by body forces or through the motion of the system boundaries. Shear work, as given by the product of the shear stress and the velocity gradient, performed on the system converts some of the kinetic energy of the mean motion into granular temperature which is the kinetic energy associated with the random particle velocities. Finally, the inelastic collisions between particles will dissipate away the granular temperature into thermodynamic heat. The magnitude of the granular temperature thus depends on the ratio of the temperature generation by shear work and its dissipation by collisional inelasticity. To close the granular temperature equation the granular conductivity, granular dissipation and the mean granular stress must be related by constitutive assumptions to the mean density, mean velocity and the granular temperature in order to complete the theory.

Lun et al. (1984) proposed a generalized model incorporating the kinetic stresses and the kinetic fluctuation-energy flux in addition to the collisional stresses, the collisional fluctuation-energy flux and the collision rate of dissipation per unit volume in the classical dense flow formulation. The constitutive relations were deduced assuming that the spherical particles are uniform, smooth, but slightly inelastic requiring that the coefficient of restitution is about unity. In order to calculate the unknown fluxes by the concept of kinetic theory, the collisional single distribution function was approximated using a scheme inspired by the Chapman-Enskog expansion method (Chapman and Cowling 1961). In this approach the parameters in the trail function were calculated by satisfying a few moment equations generated from an extended form of the Maxwell equation instead of using a Boltzmann type of equation.

Ding and Gidaspow (1990) derived a two-phase flow model starting with the Boltzmann equation for the distribution function of particles and incorporated fluid-particle interactions into the macroscopic equations. The governing equations were derived using the classical concepts of kinetic theory. The resulting macroscopic equations contain both kinetic and collisional pressures but only the collisional deviatoric stresses. The model is thus primarily intended for dense particle flows. Gidaspow (1994) proposed a modified set of macroscopic conservation equations supposedly applicable for describing both dilute and dense granular flows. Many models for granular flow were subsequently developed based on the kinetic theory approach.

Jenkins and Richman (1985) showed that a method of moments for determining the single particle distribution function of elastic molecular particles in dilute gases might be adapted and applied to dense systems of inelastic particles. By use of this method they extended the theory of Jenkins and Savage (1983) and determined a supposedly more accurate form of the balance laws including the collisional transfers and productions of the velocity moments. Strumendo et al. (2005) derived a method of moment that account for the anisotropy based on a generalization of the Chapman-Enskog expansion, and apply the method of moments to predict the behavior of fast flow of smooth, identical spheres in a circulating fluidized bed. A second-order moment method is proposed by Dan et al. (2009) based on the kinetic theory of granular flow. The transport equations for the velocity moments are presented for the particle phase. Flow behavior of gas and particles is simulated by means of gas–solid two-fluid model with the second-order moment model of particles in the bubbling fluidized bed. They find that the simulated second-order moment and normal Reynolds stresses in the vertical direction is on average 1.5–2.3 and 4.5–6.0 times larger than that in the lateral direction in a bubbling fluidized bed (Huang et al. 2013). With the kinetic theory of granular flow, transport equations for the velocity moments and closure equations are derived for the particle phase. A third-order moment model is proposed with the consideration of the increase of the binary collision probability through the correlation of particle velocity. The model for second-order moments is almost free from empirical constants, only the restitution coefficient has to be given. The anisotropic flow of particles was simulated using the method of moments (Juhui et al. 2012).

Most common particles are frictional as well as inelastic. As a result, particles can rotate with angular velocity and translate under rapid rates of deformation. During a collision of rough particles, the fluctuation energy is dissipated from inelasticity and frictions of particles. The frictional particle collision also results in the particle rotation which gives additional loss of the energy. Lun et al. (1987) developed a kinetic theory for a system of inelastic, rough spheres to study the effects of particle surface friction and rotational inertia. Jenkins and Richman (1985) used Grad's method of moments (Grad 1949) to derive conservation laws and constitutive relations for planar flows of a dense gas consisting of identical, rough, inelastic, circular disks. In these theories (Lun and Savage 1987; Jenkins and Richman 1985), two granular temperatures are involved. The first is translational granular temperature, which measures the energy associated with the translational velocity fluctuations. The second is the rotational granular temperature, which measures the energy associated with the angular velocity fluctuations. Collisional motion of rough inelastic spheres was analyzed on the basis of the kinetic Boltzmann-Enskog equation, possessing constant roughness and inelasticity. A kinetic theory of rough spheres was proposed to include rotation of particles with unequal masses and diameters (Songprawat and Gidaspow 2010). Inelastic binary collisions of particles with normal and tangential restitution coefficients are considered. New expressions for number of binary collisions, viscosities and conductivities were developed. Collision integrals produced new expressions for energy dissipation involving tangential and normal restitution coefficients. A kinetic theory for flow of dense, slightly inelastic, slightly rough sphere was proposed to study the flow behavior of gas and particles in a bubbling fluidized bed (Shuai et al. 2012a). The particle average fluctuation kinetic energy is introduced to govern the mechanism dominating kinetic energy transformation in flow of particles. The conservation equation of fluctuating kinetic energy is proposed to take the transfer of particle kinetic energy between the rotational and translational degrees of freedom and also the energy losses into account (Shuai et al. 2012b).

## 2.2 Elementary Kinetic Theory

### 2.2.1 Dynamics of an Encounter Between Two Particles

The frequency distribution of velocities of particles, $f$, is a function of position, $\mathbf{r}$ and the instantaneous velocity, $\mathbf{c}$, as well as time, $t$.

$$f = f(t, \mathbf{r}, \mathbf{c}) \tag{2.1}$$

The six coordinates, the position, r, and the velocity, c, are sufficient to determine the location of a particle, because Newton's second law has six integration constants. The velocity distribution, $f$, of the particles is often close to the normal distribution, called the Maxwellian distribution in the kinetic theory of gases. For a Maxwellian distribution, the kinetic viscosity is zero. Particle viscosity is nonzero due to particle interaction and a non-Maxwellian distribution.

The number of particles per unit volume, $n$ is given by the integral over the velocity space, $\mathbf{c}$ as

$$n = \int f d\mathbf{c} \tag{2.2}$$

The mean values of a quantity $\phi$, such as mass, momentum, energy, stress is defined to be in the usual way, as

$$n\langle\phi\rangle = \int \phi f d\mathbf{c} \tag{2.3}$$

Hence the hydrodynamic velocity, $\boldsymbol{v}$, is the integral over all the velocity space, as shown below:

$$\mathbf{v} = \frac{1}{n}\int \mathbf{c} f(\mathbf{c}) d\mathbf{c} \tag{2.4}$$

The following analysis differs from the classical kinetic theory by the introduction of a restitution coefficient first introduced, by Jenkins and Savage (1983) into the kinetic theory of granular flow. As in classical theory, consider the dynamics of an encounter of two particles of mass $m_1$ and $m_2$. Let $\mathbf{r}_1$ and $\mathbf{r}_2$ be vectors in the rest frame from a fixed origin to the two particles. Introduce the vector $\mathbf{r}_c$ from the origin to the center of mass of particles:

$$\mathbf{r}_c = \frac{m_1\mathbf{r}_1 + m_2\mathbf{r}_2}{m_1 + m_2} \tag{2.5}$$

The equations of motion for the particles are

$$m_1\frac{d^2\mathbf{r}_1}{dt^2} = \mathbf{F}_1(r) \tag{2.6}$$

$$m_2\frac{d^2\mathbf{r}_2}{dt^2} = -\mathbf{F}_1(r) \tag{2.7}$$

where the law of action and reaction has been used. On differentiation of Eq. (2.5) twice with respect to time and using Eqs. (2.6) and (2.7), the result is

$$(m_1 + m_2)^{-1}\left(m_1\frac{d^2\mathbf{r}_1}{dt^2} + m_2\frac{d^2\mathbf{r}_2}{dt^2}\right) = 0 \tag{2.8}$$

Since Eq. (2.8) shows that the acceleration of the center of mass vanishes, the center of mass is in uniform rectilinear motion. As in Chapman and Cowling (1961), let $\mathbf{G}$ be this center of mass velocity and $m_0$ be the sum of $m_1$ and $m_2$. If the velocities of particles before collision are $\mathbf{c}_1$ and $\mathbf{c}_2$ and after collision $\mathbf{c}_1'$ and $\mathbf{c}_2'$ then conservation of momentum shows that

$$m_0\mathbf{G} = m_1\mathbf{c}_1 + m_2\mathbf{c}_2 = m_1\mathbf{c}_1' + m_2\mathbf{c}_2' \tag{2.9}$$

Let $\mathbf{c}_{12}$, $\mathbf{c}'_{12}$ be the relative velocities of particles. The relative velocity of the particles after collision is

$$\mathbf{c}'_{12} = \mathbf{c}'_1 - \mathbf{c}'_2 \tag{2.10}$$

Let $\mathbf{k}$ be a unit vector directed from the center of the first particle to that of the second upon contact. Consider the particles to be smooth but inelastic with a restitution coefficient $e$ which ranges between zero and one, depending upon the material. When $e$ equals one, the relative velocities of the center of the particles are reversed upon collision, as described in Chapman and Cowling (1961). For inelastic particles, Jenkins and Savage (1983) let

$$\mathbf{k} \cdot \mathbf{c}'_{12} = -e(\mathbf{k} \cdot \mathbf{c}_{12}) \tag{2.11}$$

The individual particle velocities can be expressed in terms of the relative and the center of mass velocity as in Chapman and Cowling (1961) by solving Eq. (2.9) together with the definition of the relative velocity. For example, to obtain $\mathbf{c}_1$ and $\mathbf{c}_2$, solution of the equations

$$m_1\mathbf{c}_1 + m_2\mathbf{c}_2 = m_0\mathbf{G} \tag{2.12}$$

$$\mathbf{c}_1 - \mathbf{c}_2 = \mathbf{c}_{12} \tag{2.13}$$

obtained by determinants gives

$$\mathbf{c}_1 = \mathbf{G} + (m_2/m_0)\mathbf{c}_{12} \tag{2.14}$$

$$\mathbf{c}_2 = \mathbf{G} - (m_1/m_0)\mathbf{c}_{12} \tag{2.15}$$

Similarly, using Eqs. (2.9) and (2.10)

$$\mathbf{c}'_1 = \mathbf{G} + (m_2/m_0)\mathbf{c}'_{12} \tag{2.16}$$

$$\mathbf{c}'_2 = \mathbf{G} - (m_1/m_0)\mathbf{c}'_{12} \tag{2.17}$$

Using Eqs. (2.14) and (2.15) it can be shown that the sum of the kinetic energies of the particles is as follows:

$$\frac{1}{2}\left(m_1\mathbf{c}_1^2 + m_2\mathbf{c}_2^2\right) = \frac{1}{2}m_0\left(\mathbf{G}^2 + m_1m_2/m_0^2\mathbf{c}_{12}^2\right) \tag{2.18}$$

Similar, Eqs. (2.16) and (2.17) give

$$\frac{1}{2}\left(m_1\mathbf{c}_1'^2 + m_2\mathbf{c}_2'^2\right) = \frac{1}{2}m_0\left(\mathbf{G}^2 + m_1m_2/m_0^2\mathbf{c}_{12}'^2\right) \tag{2.19}$$

Then the translational kinetic energy change during collision $\Delta E$ is

$$2\Delta E = m_1\mathbf{c}_1'^2 + m_2\mathbf{c}_2'^2 - m_1\mathbf{c}_1^2 - m_2\mathbf{c}_2^2 = \frac{m_1m_2}{m_0}\left(\mathbf{c}_{12}'^2 - \mathbf{c}_{12}^2\right) \tag{2.20}$$

since the center of mass velocity of the two particles stays the same. Then using Eq. (2.11),

$$\Delta E = \frac{1}{2}\frac{m_1m_2}{m_0}\left(e^2 - 1\right)(\mathbf{k}\cdot\mathbf{c}_{12})^2 \tag{2.21}$$

For the case of equal mass particles,

$$m_1 = m_2 = m \tag{2.22}$$

Equation (2.21) is written as

$$\Delta E = \frac{1}{4}m\left(e^2 - 1\right)(\mathbf{k}\cdot\mathbf{c}_{12})^2 \tag{2.23}$$

Two other relations will be used in further development of the theory. Using Eqs. (2.14), (2.16) and then (2.11), it follows that

$$\mathbf{c}_1' - \mathbf{c}_1 = m_2/m_0\left(\mathbf{c}_{12}' - \mathbf{c}_{12}\right) = -m_2/m_0(1+e)(\mathbf{k}\cdot\mathbf{c}_{12})\mathbf{k} \tag{2.24}$$

which for equal masses becomes

$$\mathbf{c}_1' - \mathbf{c}_1 = -\frac{1}{2}(1+e)(\mathbf{k}\cdot\mathbf{c}_{12})\mathbf{k} \tag{2.25}$$

Similarly, using Eqs. (2.15) and (2.17) and then (2.11) for equal masses, it follows that

$$\mathbf{c}_2' - \mathbf{c}_2 = \frac{1}{2}(1+e)(\mathbf{k}\cdot\mathbf{c}_{12})\mathbf{k} \tag{2.26}$$

### 2.2.2 *Peculiar Velocity and Transport*

The transport of a quantity $\phi$, such as heat, must be invariant under a change of frame. Hence it cannot be a function of the velocity, $\boldsymbol{c}$. Otherwise it will have different values in different frames of reference. But **c**-$v$ is independent of the frame of reference. Hence, we define the difference between the instantaneous and the hydrodynamic velocities **C**, as

$$\mathbf{C} = \mathbf{c} - \mathbf{v} \tag{2.27}$$

In the kinetic theory of gases (Chapman and Cowling 1961) this difference is called the peculiar velocity. Its mean is zero, as shown below, since the mean of **c** is $v$:

$$< \mathbf{C} >=< \mathbf{c} - \mathbf{v} >= \mathbf{v} - \mathbf{v} = \mathbf{0} \tag{2.28}$$

This property is the same as that of the turbulent velocity, $v$ defined as the instantaneous minus the time average velocity.

The flux vector of $\phi$ is defined as n $< \mathbf{C}\phi(\mathbf{C}) >$. For example, if $\phi$ = E, the internal energy, then the conduction flux, q, becomes $\mathbf{q} = n < \mathrm{E}\mathbf{C} >$.

Since momentum is the mass, $m$, times the velocity, **C**, the kinetic stress tensor, $\mathbf{P}_k$ is as follows

$$\mathbf{P}_k = n < \mathbf{C}m\mathbf{C} >= \rho < \mathbf{CC} > \tag{2.29}$$

## 2.2.3 *Granular Temperature and the Equation of State*

The granular temperature is defined as the random kinetic energy of the particles. Let $\theta$ be the granular temperature be given by the random kinetic energy per unit mass as

$$\theta = 1/3\langle \mathbf{C}^2 \rangle = 1/(3n) \int_{-\infty}^{\infty}\int_{-\infty}^{\infty}\int_{x}^{\infty} (C_x^2 + C_y^2 + C_z^2) dC_x dC_y dC_z \tag{2.30}$$

in three dimensions. In two dimensions we would have only two random velocities and we would divide $< \mathbf{C}^2 >$ by two. In one dimension we have only one random velocity and the granular temperature is then simply the variance of the measured instantaneous velocities. However, its behavior is not the same as the three-dimensional granular temperature given by Eq. (2.30). The units of the granular temperature are $(\mathrm{m/s})^2$. These units are found to be convenient. The alternate definition with mass multiplying $\theta$ is not that convenient for a single particle-size mixture. It may, however, be useful for a mixture of particles (Gidaspow 1994).

The equation of state for particles is obtained by eliminating $< \mathbf{C}^2 >$ between Eqs. (2.29) and (2.30) to give

$$p_s = nm\theta \tag{2.31}$$

where the subscript, *s*, was added to emphasize that it is the solids pressure. The factor $nm$ is the bulk density. In terms of the volume fraction of solids, $\alpha_s$, and the solids density, $\rho_s$, the ideal equation of state for particles becomes as follows:

$$p_s = \alpha_s \rho_s \theta \tag{2.32}$$

## 2.3 Frequency of Collisions

The Maxwellian distribution using the granular temperature can be shown to be as follows:

$$f(\mathbf{r}, \mathbf{c}) = \frac{n}{(2\pi\theta)^{3/2}} \exp\left(-\frac{(c - v)^2}{2\theta}\right) \tag{2.33}$$

Figure 2.1 shows the geometry of collisions and the spherical coordinates and solid angle. Analogously to the single frequency distribution given by Eq. (2.33), a collisional pair distribution function, $f^{(2)}$ is introduced:

$$f^{(2)} = f^{(2)}(\mathbf{c}_1, \mathbf{r}_1, \mathbf{c}_2, \mathbf{r}_2) \tag{2.34}$$

It is defined such that

$$f^{(2)} d\mathbf{c}_1 d\mathbf{c}_2 d\mathbf{r}_1 d\mathbf{r}_2 \tag{2.35}$$

is the probability of finding a pair of particles in the volume $d\mathbf{r}_1 d\mathbf{r}_2$ centered on points $\mathbf{r}_1$, $\mathbf{r}_2$ and having velocities within the ranges $\mathbf{c}_1$ and $\mathbf{c}_1 + d\mathbf{c}_1$ and $\mathbf{c}_2$ and $\mathbf{c}_1 + d\mathbf{c}_2$.

The number of binary collisions per unit time per unit volume $N_{12}$ can be expressed as follows:

$$N_{12} = \iiint f^{(2)}(\mathbf{c}_1, \mathbf{r}, \mathbf{r} + d_{12}\mathbf{k}, \mathbf{c}_2) \cdot \mathbf{c}_{21} \cdot \mathbf{k} d_{12}^2 d\mathbf{k} d\mathbf{c}_1 d\mathbf{c}_2 \tag{2.36}$$

where

$$\mathbf{c}_{21} \cdot \mathbf{k} d\mathbf{k} = c_{21} \cos\theta \sin\theta d\theta d\phi \tag{2.37}$$

since the volume of the collision cylinder is

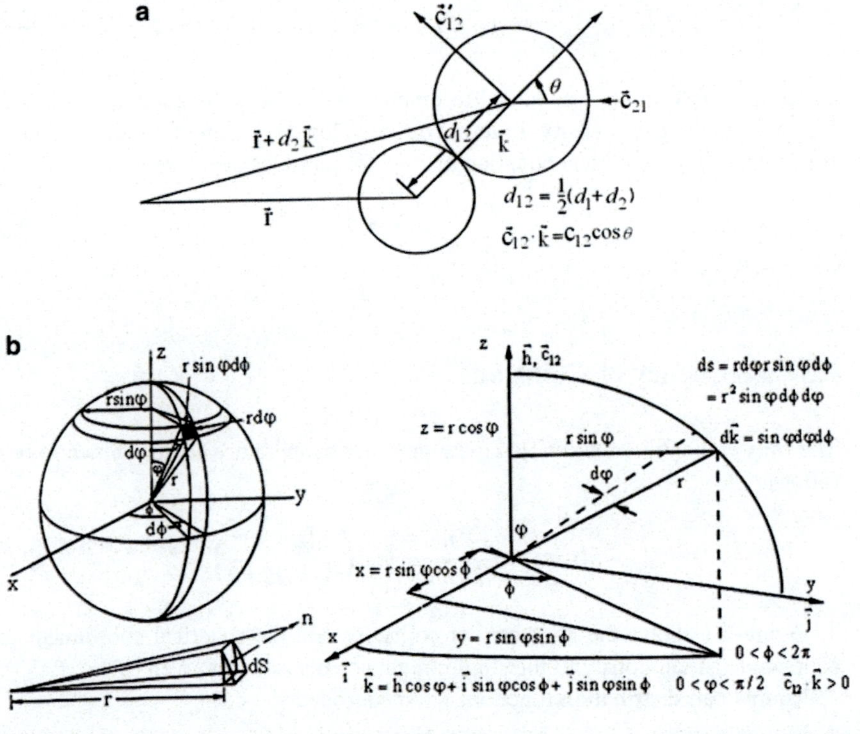


**Fig. 2.1** **a** Geometry of a collision of two spheres of diameter, $d_1$ and $d_2$. **b** Spherical coordinates and solid angle, dk.

$$d\mathbf{r} = d_{12}^2 \sin\theta d\theta d\phi \cdot c_{21} \cos\theta \cdot dt \tag{2.38}$$

The pair distribution function is expressed by a volume correction function, and Chapman and Cowling (1961) treat dense gases by assuming that

$$f^{(2)}(\mathbf{c}_1, \mathbf{r}, \mathbf{r} + d_{12}\mathbf{k}, \mathbf{c}_2) = \chi\left(\mathbf{r} + \frac{1}{2}d_{12}\mathbf{k}\right) \cdot f(\mathbf{r}, \mathbf{c}_1) \cdot f(\mathbf{r} + d_{12}\mathbf{k}, \mathbf{c}_2) \tag{2.39}$$

where $\chi$ is a quantity that is unity for ordinary gases, and increases with increasing density, becoming infinite as the gas approaches the state in which the molecules are packed so closely that motion is impossible. The position of the point of collision can be expressed in different ways depending on the choice of origin, normally the center point of particle 1 is used.

The dependence on position indicated in Eq. (2.39) can be related to the volume fraction of solid. A form successfully used by Ding and Gidaspow (1990) is

$$\chi\left(\mathbf{r}+\frac{1}{2}d_{12}\mathbf{k}\right)=g_0(\alpha_s)=\left[1-\left(\frac{\alpha_s}{\alpha_{s,\max}}\right)^{1/3}\right]^{-1} \tag{2.40}$$

The radial distribution function, $g_O \geq 1$, accounts for the increase of the binary collision probability when the suspension becomes denser. This form requires, as an empirical input, the maximum packed solids fraction, $\alpha_{s,\max}$.Then $f^{(2)}$ becomes

$$f^{(2)}=g_0\frac{n_1n_2}{(2\pi\theta)^3}\exp\left[-\frac{(\mathbf{c}_1-\mathbf{v})^2+(\mathbf{c}_2-\mathbf{v})^2}{2\theta}\right] \tag{2.41}$$

For a uniform, steady state, **v** is zero. The integrals over $k$ as seen from Eq. (2.37) give

$$\int_0^{\pi}\cos\theta\,\sin\theta d\theta=\frac{1}{2} \tag{2.42a}$$

and

$$\int_0^{2\pi}d\phi=2\pi \tag{2.42b}$$

Thus, the number of collisions per unit time per unit volume using (2.40) to (2.42) in Eq. (2.36) is as follows:

$$N_{12}=\frac{\pi d_{12}^2g_0n_1n_2}{(2\pi\theta)^3}\iint c_{21}\exp\left[-\frac{c_1^2+c_2^2}{2\theta}\right]d\mathbf{c}_1d\mathbf{c}_2 \tag{2.43}$$

For $m_1=m_2$, Eq. (2.18) becomes

$$c_1^2+c_2^2=2G^2+\frac{1}{2}c_{12}^2 \tag{2.44}$$

In Eq. (2.43) the coordinates are transformed from $\mathbf{c}_1,\mathbf{c}_2$ to $\mathbf{G},\mathbf{c}_{12}$. As in Chapman and Cowling (1961), the Jacobian of transformation using Eq. (2.14) is

$$\frac{\partial(\mathbf{G},\mathbf{c}_{12})}{\partial(\mathbf{c}_1,\mathbf{c}_2)}=\frac{\partial(\mathbf{c}_1+m_2/m_0\mathbf{c}_{21},\mathbf{c}_2-\mathbf{c}_1)}{\partial(\mathbf{c}_1,\mathbf{c}_2)}=\frac{\partial(\mathbf{c}_1,\mathbf{c}_2)}{\partial(\mathbf{c}_1,\mathbf{c}_2)}=1 \tag{2.45}$$

From spherical symmetry,

$$d\mathbf{G}=4\pi G^2dG \tag{2.46}$$

and

$$d\mathbf{c}_{21} = 4\pi c_{21}^2 dc_{21} \tag{2.47}$$

Hence, using (2.44) through (2.47), Eq. (2.43) becomes

$$N_{12} = \frac{2n_1 n_2 d_{12}^2 g_0}{\theta^3} \int_0^\infty \int_0^\infty \exp\left[ -\frac{2G^2 + \frac{1}{2}c_{21}^2}{2\theta} \right] G^2 dG c_{21}^3 dc_{21} \tag{2.48}$$

Collecting all terms, the collision frequency becomes

$$N_{12} = 4n_1 n_2 d_{12}^2 g_0 \sqrt{\pi\theta} \tag{2.49}$$

In Eq. (2.49) as $g_0$ becomes infinite at maximum packing, the number of collisions becomes infinite, but at this point the granular temperature is expected to approach zero, leaving $N_{12}$ undefined.

## 2.4 Boltzmann Integral—Differential Equation

The Boltzmann equation for the frequency distribution $f$, Gidaspow (1994) can be derived, as any conversation equation, using the Reynolds transport theorem. The conversation principle is as follows:

$$\frac{d}{dt} \int_{\mathbf{r}} \int_{\mathbf{c}} f(t, \mathbf{r}, \mathbf{c}) d\mathbf{r} d\mathbf{c} = \int_{\mathbf{r}} \int_{\mathbf{c}} \left( \frac{\partial f}{\partial c} \right)_{coll} d\mathbf{r} d\mathbf{c} \tag{2.50}$$

Then an application of the usual Reynolds transport theorem, it gives the well-known Boltzmann equation

$$\frac{\partial f}{\partial t} + \mathbf{c} \cdot \frac{\partial f}{\partial \mathbf{r}} + \mathbf{F} \frac{\partial f}{\partial \mathbf{c}} = \left( \frac{\partial f}{\partial t} \right)_{coll} \tag{2.51}$$

where $\mathbf{c}$ and $\mathbf{r}$ were regarded to be as independent coordinates and where Newton's law of motion was

$$\frac{Force}{Unit\,mass} \equiv \mathbf{F} = \frac{d\mathbf{c}}{dt} \tag{2.52}$$

For binary collisions of rigid particles, the right-hand side of the Boltzmann Eq. (2.51) assumes the form

$$\left(\frac{\partial f}{\partial t}\right)_{coll} = \iint \left(f^{(2)}\mathbf{c}'_{12}\cdot\mathbf{k} - f^{(2)}\mathbf{c}_{12}\cdot\mathbf{k}\right)d_{12}^2 d\mathbf{k}d\mathbf{c}_1 \tag{2.53}$$

where the primes again indicate the quantities after collision and $f^{(2)}$ is the product of the respective single particle distributions as given by Eq. (2.39). Hence, the Boltzmann equation is an integral–differential equation. Because of its non-linearity it must be solved by iteration. For the first approximation, one takes the Maxwellian distribution. The second approximation will give rise to a Navier–Stokes type equation. This is done efficiently using an altered form of the Boltzmann operator, as presented below.

$$\mathbf{C} = \mathbf{c} - \mathbf{v}(t, \mathbf{r}) \tag{2.54}$$

changes the coordinates from $\mathbf{c}$ to $\mathbf{C}$. Then

$$f(t, \mathbf{r}, \mathbf{c}) = f_C(t, \mathbf{r}, \mathbf{C}) \tag{2.55}$$

By chain rules of calculus,

$$\frac{\partial f}{\partial \mathbf{c}} = \frac{\partial f_C}{\partial \mathbf{C}} \tag{2.56}$$

$$\frac{\partial f}{\partial t} = \frac{\partial f_C}{\partial t} - \frac{\partial f_C}{\partial \mathbf{C}}\frac{\partial \mathbf{v}}{\partial t} \tag{2.57}$$

$$\frac{\partial f}{\partial x} = \frac{\partial f_C}{\partial x} - \frac{\partial f_C}{\partial \mathbf{C}}\frac{\partial \mathbf{v}}{\partial x} \tag{2.58}$$

Differentiation gives

$$\frac{\partial f}{\partial \mathbf{r}} = \frac{\partial f_C}{\partial \mathbf{r}} - \frac{\partial f_C}{\partial \mathbf{C}}\frac{\partial \mathbf{v}}{\partial \mathbf{r}} \tag{2.59}$$

Then, multiplication of Eq. (2.59) by $\mathbf{C}$ produces the double dot sum useful in the derivation of Navier–Stokes equation:

$$\mathbf{C}\cdot\frac{\partial f}{\partial \mathbf{r}} = \mathbf{C}\cdot\frac{\partial f_C}{\partial \mathbf{r}} - \frac{\partial f_C}{\partial \mathbf{C}}\left(\mathbf{C}\frac{\partial}{\partial \mathbf{r}}\right)\mathbf{v} \tag{2.60}$$

Using the notation of Chapman and Cowling (1961) for the convective derivative following the hydrodynamic velocity,

$$\frac{D}{Dt} = \frac{\partial}{\partial t} + \mathbf{v}\frac{\partial}{\partial \mathbf{r}} \tag{2.61}$$

the Bolzmann equation takes the modified form

$$\frac{Df}{Dt} + \mathbf{C}\frac{\partial f}{\partial \mathbf{r}} + \left(\mathbf{F} - \frac{D\mathbf{v}}{Dt}\right)\frac{\partial f}{\partial \mathbf{C}} - \frac{\partial f}{\partial \mathbf{C}}\mathbf{C} : \frac{\partial \mathbf{v}}{\partial \mathbf{r}} = \left(\frac{\partial f}{\partial t}\right)_{coll} \tag{2.62}$$

## 2.5 Kinetic Theory of Granular Flow

### 2.5.1 *Maxwell's Transport Equation*

A transport equation for a quantity $\psi$ can be obtained starting with the Boltzman equation by multiplying it by $\psi$ and integrating over $\mathbf{c}$ as shown below:

$$\int \psi\left(\frac{\partial f}{\partial t} + \mathbf{c}\cdot\frac{\partial f}{\partial \mathbf{r}} + \mathbf{F}\cdot\frac{\partial f}{\partial \mathbf{c}}\right)d\mathbf{c} = \int \psi\left(\frac{\partial f}{\partial t}\right)_{coll} d\mathbf{c} \tag{2.63}$$

$$\int \psi\frac{\partial f}{\partial t}d\mathbf{c} = \frac{\partial}{\partial t}\int \psi f d\mathbf{c} - \int f\frac{\partial \psi}{\partial t}d\mathbf{c} = \frac{\partial n\langle\psi\rangle}{\partial t} - n\left\langle\frac{\partial \psi}{\partial t}\right\rangle \tag{2.64a}$$

$$\int \psi c_x\frac{\partial f}{\partial t}d\mathbf{c} = \frac{\partial}{\partial t}\int \psi c_x f d\mathbf{c} - \int f c_x\frac{\partial \psi}{\partial t}d\mathbf{c} = \frac{\partial n\langle\psi c_x\rangle}{\partial t} - n\left\langle c_x\frac{\partial \psi}{\partial t}\right\rangle \tag{2.64b}$$

$$\int \psi\frac{\partial f}{\partial t}d\mathbf{c} = \iint [\psi f]_{c_x=-\infty}^{c_x=\infty}dc_y dc_x - \int f\frac{\partial \psi}{\partial c_x}d\mathbf{c} = -n\left\langle\frac{\partial \psi}{\partial c_x}\right\rangle \tag{2.64c}$$

In (2.64b) as $\mathbf{c}$ is independent of $\mathbf{r}$, the variable $c_x$ is not included in the differentiation with respect to $x$; in (2.64c), an integration by parts is performed, and the integrated part vanishes because, by hypothesis, $\psi f$ tends to zero as $c_x$ tends to infinity in either direction.

In this way it is found that

$$\frac{\partial n\langle\psi\rangle}{\partial t} + \frac{\partial}{\partial \mathbf{r}}\cdot n\langle\psi\mathbf{c}\rangle - n\left[\left\langle\frac{\partial \psi}{\partial t}\right\rangle + \left\langle\mathbf{c}\cdot\frac{\partial \psi}{\partial \mathbf{r}}\right\rangle + \mathbf{F}\cdot\left\langle\frac{\partial \psi}{\partial \mathbf{c}}\right\rangle\right] = \int \psi\left(\frac{\partial f}{\partial t}\right)_{coll} d\mathbf{c} \tag{2.65}$$

This equation can be again written in a form more convenient for some applications by changing $\psi$ into the peculiar velocity $\mathbf{C}$ dependence. Using the chain rules for $\psi$ as given by Eqs. (2.56) to (2.59), it becomes as follows for $\psi = \psi\langle\mathbf{C}\rangle$ only:

$$\frac{Dn\langle\psi\rangle}{Dt} + n\langle\psi\rangle\frac{\partial \mathbf{v}}{\partial \mathbf{r}} + \frac{\partial}{\partial \mathbf{r}}n\langle\psi\mathbf{C}\rangle - n\left(\mathbf{F} - \frac{D\mathbf{v}}{Dt}\right)\left\langle\frac{\partial \psi}{\partial \mathbf{C}}\right\rangle - n\left\langle\frac{\partial \psi}{\partial \mathbf{C}}\mathbf{C}\right\rangle : \frac{\partial \mathbf{v}}{\partial \mathbf{r}} = n\psi_C \tag{2.66}$$

in which $\psi_C$ denotes the collisional rate of change of the macroscopic variable $\langle\psi\rangle$. The values of $f^{(2)}$ separated by a distance $d_{12}$k and are expanded by means of the Taylor series

$$f^{(2)}(t,\mathbf{r},\mathbf{c}_1,\mathbf{r}+d_{12}\mathbf{k},\mathbf{c}_2)=f^{(2)}(t,\mathbf{r}-d_{12}\mathbf{k},\mathbf{c}_1,\mathbf{r},\mathbf{c}_2)$$
$$+(d_{12}\mathbf{k}\cdot\nabla)\sum_{m=0}^{\propto}\frac{(-d_{12}\mathbf{k}\cdot\nabla)^m}{(m+1)!}f^{(2)}(t,\mathbf{r},\mathbf{c}_1,\mathbf{r}+d_{12}\mathbf{k},\mathbf{c}_2) \quad (2.67)$$

By inserting $f^{(2)}(t,\mathbf{r},\mathbf{c}_1,\mathbf{r}+d_{12}\mathbf{k},\mathbf{c}_2)$, adding the collisional rate term, we obtain

$$\Delta(\psi)_{\mathrm{col}}=\Delta(\psi_1)_{\mathrm{col}}+\Delta(\psi_2)_{\mathrm{col}}=\Omega(\psi)-\nabla\cdot\Phi(\psi) \quad (2.68)$$

The source term $\Omega(\psi)$ represents the loss of $\psi$ caused by inelastic collisions. It is defined by:

$$\Omega(\psi)=\frac{d_p^2}{2}\int\limits_{\mathbf{g}_{21}\cdot\mathbf{k}>0}[(\psi_2'+\psi_1')-(\psi_2+\psi_1)]f^{(2)}(\mathbf{r}-d_{12}\mathbf{k},\mathbf{c}_1,\mathbf{r},\mathbf{c}_2,t)(\mathbf{g}_{21}\cdot\mathbf{k})d\mathbf{k}d\mathbf{c}_1d\mathbf{c}_2 \quad (2.69)$$

The flux term represents the transfer of $\psi$ during collisions. It is defined by

$$\Phi(\psi)=-\frac{d_p^3}{2}\int\limits_{\mathbf{g}_{21}\cdot\mathbf{k}>0}(\psi_1'-\psi_1)\mathbf{k}\sum_{m=0}^{\propto}\frac{(-d_{12}\mathbf{k}\cdot\nabla)^m}{(m+1)!}f^{(2)}(\mathbf{r},\mathbf{c}_1,\mathbf{r}+d_{12}\mathbf{k},\mathbf{c}_2,t)(\mathbf{g}_{21}\cdot\mathbf{k})d\mathbf{k}d\mathbf{c}_1d\mathbf{c}_2$$
$$\approx-\frac{d_p^3}{2}\int\limits_{\mathbf{g}_{21}\cdot\mathbf{k}>0}(\psi_1'-\psi_1)\mathbf{k}f^{(2)}(\mathbf{r},\mathbf{c}_1,\mathbf{r}+d_{12}\mathbf{k},\mathbf{c}_2,t)(\mathbf{g}_{21}\cdot\mathbf{k})d\mathbf{k}d\mathbf{c}_1d\mathbf{c}_2 \quad (2.70)$$

This relation represents a first order approximation to the flux since only the first term in the Taylor series is included. To calculate the integrals defining the source term and the flux term, appropriate expressions for $\Delta\psi$ and $\psi_1'-\psi_1$ have to be determined from an analysis of the inelastic binary particle collision dynamics.

Let $\psi=m$, the conservation equation of particles is shown below:

$$\frac{D(nm)}{Dt}+nm\frac{\partial\mathbf{v}_s}{\partial\mathbf{r}}=0 \quad (2.71)$$

Since $\rho_s\alpha_s=nm=\rho$, Eq. (2.71) is the standard continuity equation with no phase change:

$$\frac{\partial(\alpha_s\rho_s)}{\partial t}+\frac{\partial(\alpha_s\rho_s\mathbf{v}_s)}{\partial\mathbf{r}}=0 \quad (2.72)$$

The subscript $s$ was added to emphasize that the velocity in Eq. (2.72) is the solid velocity.

Let $\psi = m\mathbf{C}$, the relative velocity of particles relative to its hydrodynamics velocity. It is clear that

$$\langle \mathbf{C} \rangle = \langle \mathbf{c} \rangle - \mathbf{v} = 0 \tag{2.73}$$

The last term in Eq. (2.66) gives

$$\left\langle \frac{\partial m\mathbf{C}}{\partial \mathbf{C}} \mathbf{C} \right\rangle = m\langle \mathbf{C} \rangle = 0 \tag{2.74}$$

The force $\mathbf{F}$ in Eq. (2.66) consists of the externally applied force and the force of interaction between phases. The latter force consists of the drag between the phases and other forces such as the added mass force for acceleration of a gas bubble injected into a liquid. Clearly, Eq. (2.66) does not include the static particle-to-particle stress. The gradient of this solids stress must be either added to Eq. (2.66) or the kinetic theory approach restricted to flows with no dead regions. The former approach is clearly preferable. For dilute suspensions, fluid turbulence must also be included in force $\mathbf{F}$. For now the only effects included are the gravity, the drag and the buoyancy effect which appears through the gradient of fluid pressure. Then

$$\rho \mathbf{F} = \alpha_s \rho_s \mathbf{g} + \beta_A (\mathbf{v}_f - \mathbf{v}_s) - \alpha_s \nabla p_f \tag{2.75}$$

Hence, the momentum balance of the particles can be written as follows in

$$\frac{\partial(\alpha_s \rho_s \mathbf{v}_s)}{\partial t} + \nabla \cdot (\alpha_s \rho_s \mathbf{v}_s \mathbf{v}_s) = -\alpha_s \nabla p_f + \nabla p_s + \nabla \cdot \alpha_s \tau_s + \alpha_s \rho_s \mathbf{g} + \beta_A (\mathbf{v}_f - \mathbf{v}_s) \tag{2.76}$$

where $p_s$ and $\tau_s$ are the pressure and stress tensor of the granular material. These terms represent the transport of momentum by particle velocity fluctuations and by particle collisions, respectively.

The summation of volume fractions for all phases is equal to one.

$$\alpha_f + \alpha_s = 1 \tag{2.77}$$

Let $\psi = \frac{1}{2} m\mathbf{C}^2$, the relative kinetic energy of the particles, where $\frac{1}{3}\mathbf{C}^2$ is the granular temperature $\theta$. The granular temperature is a measure of the random particle kinetic energy per unit mass. It is produced due to "viscous-type dissipation" and consumed due to inelastic collisions. The heat flux $\mathbf{q}_K$ is defined as

$$\mathbf{q}_K = \frac{1}{2} nm \langle \mathbf{C}^2 \mathbf{C} \rangle = \frac{3}{2} n \langle \theta \mathbf{C} \rangle \tag{2.78}$$

The fourth term in Eq. (2.66) vanishes because

$$\left\langle \frac{\partial 1/2m\mathbf{C}^2}{\partial \mathbf{C}} \right\rangle = m\langle \mathbf{C} \rangle = 0 \tag{2.79}$$

The production of the fluctuations due to the force is

$$\begin{aligned} \rho < \mathbf{Fc} > -\rho \mathbf{F} \cdot \mathbf{v} &= \beta_A\big[< \mathbf{C}_p \cdot (\mathbf{C}_g - \mathbf{C}_p) > -\mathbf{v}_p \cdot (\mathbf{v}_g - \mathbf{v}_p)\big] \\ &= \beta_A < \mathbf{C}_g \cdot \mathbf{C}_p > - 3\beta_A \theta \end{aligned} \tag{2.80}$$

where $\langle C_g C_p \rangle$ is the correlation between the velocity fluctuations of the gas and the particles. Through this term, gas oscillations produce oscillations of particles. $k_f$ is the turbulent kinetic energy of the fluid. The first part is minor for laminar flows.

The final form of the granular temperature equation for the particle phase can be expressed as,

$$\begin{aligned} &\frac{3}{2}\left[\frac{\partial}{\partial t}(\alpha_s \rho_s \theta) + \nabla \cdot (\alpha_s \rho_s \theta \mathbf{v}_s)\right] \\ &= (p_s + \tau_s) : \nabla \mathbf{v}_s - \nabla(\mathbf{q}_k + \mathbf{q}_c) + N_c\left(\frac{m}{2}\mathbf{c}^2\right) + \beta_A < \mathbf{C}_g \cdot \mathbf{C}_p > -3\beta_A \theta \end{aligned} \tag{2.81}$$

On the right-hand side of Eq. (2.81), the first term represents the energy production due to the deformation work, the second term is the energy transfer. The third term is the collisional energy dissipation due to inelastic collisions, and the last two terms are the net rate of transfer of fluctuation energy between the two phases.

It is noted that the properties reflected by the granular temperature $\theta$ are quite different from those of the thermal temperature $T$. The granular temperature is not representing a noticeable sensible heat because the magnitude of the translational energy (1/2)$mC^2$ of granular flows is similar to the average translational kinetic energy (1/2)$m\mathbf{v_s}^2$. For comparison, in a gas the peculiar velocity of the molecules $C_m$ are much larger than the average gas velocity $v_c$, i.e., $C_m \gg < \mathbf{v}_c >$. However, the translational energy associated with a granular material is assumed characterized by scales comparable to the particle size being considerable smaller than the microscopic length scales of the flow. Therefore, the particle fluctuations in granular flows are considered a local property of solids phase. In this interpretation the granular temperature must not be confused with the turbulent kinetic energy which is a continuum flow phenomenon. In a second view, the particles are assumed to be characterized by scales comparable to the macroscopic length scales of the flow, thus the granular temperature is associated with the turbulent kinetic energy which is a continuum fluid flow phenomenon.

### 2.5.2 Constitutive Correlations

The stress tensor for fluid phase is given by a Newtonian-type viscous approximation, as

$$\tau_f = \alpha_f \mu_{fe}\left(\nabla \mathbf{v}_f + \nabla \mathbf{v}_f^T\right) - \frac{2}{3}\alpha_f \mu_{fe} \nabla \cdot \mathbf{v}_f \mathbf{I} \tag{2.82}$$

where the fluid phase effective viscosity $\mu_{\text{fe}}$ is calculated according to the fluid dynamic viscosity $\mu_f$ and fluid turbulent viscosity $\mu_{ft}$. The fluid phase turbulent viscosity is determined from a two equation $k_f$-$\varepsilon_f$ turbulence model.

$$\mu_{ft} = C_\mu \rho_f \frac{k_f^2}{\varepsilon_f} \tag{2.83}$$

The equations of fluid phase turbulent kinetic energy $k_f$ and its dissipation rate $\varepsilon_f$ are (Syamlal et al. 1993; ANSYS 2005; Huilin 2017):

$$\frac{\partial}{\partial t}(\alpha_f \rho_f k_f) + \nabla \cdot (\alpha_f \rho_f \mathbf{v}_f k_f) = \nabla \cdot \left(\alpha_f \frac{\mu_{ft}}{\sigma_k} \nabla k_f\right) + \alpha_f G_k + \beta_A (k_{fs} - 2k_f) - \alpha_f \rho_f \varepsilon_f \tag{2.84}$$

$$\frac{\partial}{\partial t}(\alpha_f \rho_f \varepsilon_f) + \nabla \cdot (\alpha_f \rho_f \mathbf{v}_f \varepsilon_f) = \nabla \cdot \left(\alpha_f \frac{\mu_{ft}}{\sigma_\varepsilon} \nabla \varepsilon_f\right) + \alpha_f \frac{\varepsilon_f}{k_f}(C_1 G_k - C_2 \rho_f \varepsilon_f) + C_3 \left(\frac{\varepsilon_f}{k_f}\right) \beta_A (k_{fs} - 2k_f) \tag{2.85}$$

$$k_{fs} = \frac{\eta_r}{1 + \left(1 + \frac{\alpha_s \rho_s}{\alpha_f \rho_f}\right)\eta_r}\left(2k_f + 3\frac{\alpha_s \rho_s}{\alpha_f \rho_f}\theta\right) \tag{2.86}$$

where $C_1, C_2, C_3, C_\mu, \sigma_{\text{k}}$ and $\sigma_\varepsilon$ are empirical constants, and they are 1.44, 1.92, 1.22, 0.09, 1.0 and 1.3, respectively. The last term in the right hand side of $k_g$ and $\varepsilon_g$ equations represents the fluctuation energy interaction term. They represent the kinetic energy necessary to accelerate discrete particles, or the kinetic energy exchanged from the Euler solid phase to the Euler gas phase if the Euler solid phase has large fluctuation velocities in a region where the Euler gas phase turbulent kinetic energy is small. The interaction fluctuation energy accounts for the transfer of fluctuation energy within the Euler gas phase and the Euler solid phase. The generation of turbulent kinetic energy due to the mean velocity gradients $G_{\text{k}}$ is

$$G_k = \mu_{ft} \mathbf{S}^2 \tag{2.87}$$

$$S = \sqrt{2S_{ij}S_{ij}} \quad \text{and} \quad S_{ij} = \frac{1}{2}\left(\frac{\partial \mathbf{v}_{fi}}{\partial x_j} + \frac{\partial \mathbf{v}_{fj}}{\partial x_i}\right) \tag{2.88}$$

The ratio between the Lagrangian integral time scale and the particle relaxation time:

$$\eta_{\mathrm{r}} = \frac{\tau_{fs}}{\tau_t} = \frac{3}{2}\frac{k_f}{\varepsilon_f}\frac{\beta_A C_\mu}{\alpha_s \rho_s \sqrt{1 + 0.45\frac{3|v_f - v_s|}{2k_f}}} \tag{2.89}$$

Detailed derivation for the constitutive equation in a dense system for the particle pressure; shear viscosity and bulk viscosity as a function of granular temperature, can be found elsewhere (Ding and Gidaspow 1990; Gidaspow 1994). By use of the Chapman-Enskog approximate solution method (Chapman and Cowling 1961), the kinetic pressure tensor can be given by:

$$\begin{aligned} p_{\mathrm{k}} &= \alpha_s \rho_s < \mathbf{CC} >= \alpha_s \rho_s \int \mathbf{CC} f d\mathbf{C} = \alpha_s \rho_s \int \mathbf{CC} f^0 (1 + \phi_1) d\mathbf{C} \\ &= \alpha_s (p_{kin}\mathbf{I} + \sigma_{kin}) \end{aligned} \tag{2.90}$$

where $p_{kin}$ is the kinetic pressure.

$$p_{kin} = \rho_s \theta \tag{2.91}$$

The $\sigma_{kin}$ denotes the deviatoric pressure tensor defined by:

$$\sigma_{kin} = -\frac{2}{(1+e)g_0}\left(1 + \frac{4}{5}(1+e)\alpha_s g_0\right) 2\mu_{dil}\mathrm{S} \tag{2.92}$$

Gidaspow (1994) proposed a closure for the collisional pressure tensor. By use of the Chapman-Enskog approximate solution method, the collisional pressure tensor can be written as:

$$p_c = p_{col1} + p_{col2} = \alpha_s (p_{col}\mathbf{I} + \sigma_{col}) \tag{2.93}$$

where $\mathrm{p}_{col1}$ is given by:

$$\begin{aligned} p_{col1} &= \frac{2}{5}\alpha_s^2 \rho_s g_0 (1 + \mathrm{e})(2 < \mathbf{CC} > + \mathbf{C}^2\mathbf{I}) \\ &= 2\alpha_s^2 \rho_s g_0 (1+e)\theta \mathbf{I} - \frac{2\alpha_s}{(1+e)g_0}(1 + \frac{4}{5}\alpha_s g_0 (1+e))\frac{4}{5}\alpha_s^2 \rho_s g_0 (1+e) 2\mu_{dil}\mathbf{S} \end{aligned} \tag{2.94}$$

and, the $p_{col2}$ is:

$$p_{col2} = -\frac{4}{3\sqrt{\pi}}\alpha_s^2\rho_s d_p g_0(1+e)\sqrt{\theta}\left[\frac{6}{5}\mathbf{S} + (\nabla\cdot\mathbf{v}_s)\mathbf{I}\right] \tag{2.95}$$

It is recognized that the first term in the formula for $p_{col}$ contains the collisional pressure, defined by:

$$p_{col} = 2\alpha_s^2\rho_s g_0(1+e)\theta \tag{2.96}$$

The collisional deviatoric pressure tensor consists of the remaining terms in $p_{col}$ and is thus given by:

$$\begin{aligned}\sigma_{col} = &-\frac{2\alpha_s}{(1+e)g_0}\left(1+\frac{4}{5}\alpha_s g_0(1+e)\right)\frac{4}{5}\alpha_s\rho_s g_0(1+e)2\mu_{dil}\mathbf{S}\\ &-\frac{4}{3\sqrt{\pi}}\alpha_s\rho_s d_p g_0(1+e\sqrt{\theta}\left[\frac{6}{5}\mathbf{S}+(\nabla\cdot\mathbf{v}_s)\mathbf{I}\right]\end{aligned} \tag{2.97}$$

The total pressure tensor contains both the kinetic and collisional contributions. Hence it follows that the total pressure of the dispersed phase is defined as the sum of the kinetic Eq. (2.91) and collisional Eq. (2.96) pressure contributions (Gidaspow and Huilin, 1998):

$$p_s = p_{kin} + p_{col} = \varepsilon_s\rho_s\theta + 2\rho_s(1+e_s)\varepsilon_s^2 g_o\theta \tag{2.98}$$

The solid stress tensor can be written as:

$$\tau_s = \alpha_s\mu_s(\nabla\mathbf{v}_s + \nabla\mathbf{v}_s^T) + \alpha_s(\xi_s - \frac{2}{3}\mu_s)\nabla\cdot\mathbf{v}_s\mathbf{I} \tag{2.99}$$

where the shear viscosity, $\mu_s$; and bulk viscosity, $\xi_s$, are expressed as a function of granular temperature based on the kinetic theory model (Gidaspow 1994).

$$\mu_s = \frac{2\mu_{dil}}{(1+e)\alpha_s g_o}[1+\frac{4}{5}(1+e)\alpha_s g_o]^2 + \frac{4}{5}\alpha_s\rho_s d_p g_0(1+e)\sqrt{\frac{\theta}{\pi}} \tag{2.100}$$

The first term on the RHS in the viscosity closure denotes the kinetic contribution and dominates in the dilute regime. The second term on the RHS denotes the collisional contribution and dominates in the dense flow regime. The viscosity for the dilute limit can be written as:

$$\mu_{dil} = \frac{30\sqrt{2}}{96}\rho_s\alpha_s \min\left|\frac{1}{6\sqrt{2}}\frac{d_p}{\alpha_s}, \frac{1}{6\sqrt{2}}\frac{d_p}{\alpha_{s,\min}}\right|\sqrt{\pi\theta} \tag{2.101a}$$

This relation for $\mu_{dil}$ becomes zero when $\alpha_s < \alpha_{s,\min}$. A typical value, $\alpha_{s,min} = 1 \times 10^{-5}$, is used to limit dilute flow of fluid-particles systems. It is also expressed as a function of granular temperature based on the kinetic theory model (Gidaspow 1994)

$$\mu_{dil} = \frac{5}{96}\rho_s d_p \sqrt{\pi\theta} \tag{2.101b}$$

The normal component of the collisional deviatoric pressure tensor resembles the bulk viscosity term and is thus denoted by

$$\xi_s = \frac{4}{3}\alpha_s \rho_s d_p g_0 (1+e)\sqrt{\frac{\theta}{\pi}} \tag{2.102}$$

A similar derivation can be made for the granular heat flux vectors (Gidaspow 1994). The total granular heat flux is the sum of the kinetic flux $q_k$ and the collisional flux $q_c$, and expressed by granular conductivity.

$$q_s = q_k + q_c = k_s \nabla\theta \tag{2.103}$$

The granular conductivity, $k_s$ consists of the kinetic part due to the elastic particles derived from dilute kinetic theory of gases (Chapman and Cowling 1961) and the collisional part due to the inelastic collision of particles (Gidaspow 1994):

$$k_s = \frac{150\rho_s d_p \sqrt{\pi\theta}}{384(1+e)g_0}\left[1 + \frac{6}{5}\alpha_s g_0 (1+e)\right]^2 + 2\rho_s \alpha_s^2 d_p (1+e) g_0 \sqrt{\frac{\theta}{\pi}} \tag{2.104}$$

the first term on the RHS of the conductivity closure denotes the kinetic contribution and dominates in the dilute regime. The second term on the RHS of the closure equation denotes the collisional contribution and dominates in the dense flow regime.

In this analysis, it is assumed that the interaction force between fluid and particles is due only to drag. Hence, the rate of energy dissipation per unit volume due to inelastic collisions of particles is expressed as a function of granular temperature.

$$\gamma_s = -N_c\left(\frac{m}{2}\mathrm{c}^2\right) = 3(1-e^2) g_0 \rho_s \alpha_s^2 \theta \left(\frac{4}{d_p}\sqrt{\frac{\theta}{\pi}} - \nabla \cdot \mathbf{v}_s\right) \tag{2.105}$$

## 2.6 Kinetic-Frictional Stress Models

When the solid volume fraction is high, the most common approach to consider the effect of frictional stresses is the kinetic-frictional model based on the addition of stress from the two limiting regimes at a critical solid volume fraction ($\alpha_{cr}$) (Johnson

1985; Johnson and Jackson 1987; Syamlal et al. 1993; Huilin et al. 2003; Srivastava and Sundaresan 2003; Shuyan et al. 2009; Liyan et al. 2012)

$$\tau_s = \tau_{kin} + \tau_{fri} \tag{2.106}$$

This approach is based on the assumptions that consider the solid stress comes from the kinetic and collisional stress $\tau_{kin}$, and frictional contributions $\tau_{fri}$ in an additive manner, where the frictional contributions appear only at higher solid volume fractions (i.e., greater than 0.5). Although, this approach lacks a strong physical justification and the hypothetical assumption of the critical solid volume fraction remains without experimental proof, the theory has been shown to capture the qualitative features of slow dense solid flows (Srivastava and Sundaresan 2003).

The constitutive equation for the solids pressure $p_s$ and shear viscosity $\mu_s$ consists of three sources, namely, kinetic, collision, and friction, which could be written either in an additive manner or a continuous form as

$$p_s = p_{kin} + p_{fri} = (p_k + p_c) + p_{fri} \quad \alpha_s \geq \alpha_{cr} \tag{2.107}$$

$$\mu_s = \mu_{kin} + \mu_{fri} = (\mu_k + \mu_c) + \mu_{fri} \quad \alpha_s \geq \alpha_{cr} \tag{2.108}$$

Schaeffer (1987) derived the following expression for the frictional stress, by assuming the system to be perfectly rigid-plastic, incompressible, non-cohesive, Coulomb powder with a yield surface of von Mises type, and the eigenvectors of the strain rate and stress tensors are parallel, as

$$\mu_{fri} = \frac{p_c \sin\varphi}{\sqrt{4\Pi_{2D}}} \tag{2.109}$$

where $\varphi$ is the angle of internal friction, $\text{II}_{2\text{D}}$ is the second invariant of the deviatoric stress tensor, and $p_c$ is the critical state pressure.

Johnson and Jackson (1987) proposed a critical state solid frictional pressure that allows for a slight compressibility with very limited particle concentration change. The Johnson and Jackson correlation for frictional pressure can be written as

$$p_{fri} = Fr\frac{(\alpha_s - \alpha_{\min})^q}{(\alpha_{\max} - \alpha_s)^p} \quad \alpha_{\min} = 0.5 \leq \alpha_s \leq \alpha_{\max} = 0.63 \tag{2.110}$$

where the coefficients $q$ and $p$ are to be 2.0 and 3.0. $Fr$ is a coefficient with different values reported in the literature from 0.05 to 5 (Johnson et al., 1990). The coefficient $Fr$ was modified by means of a function of the volume fraction as $Fr = 0.1\alpha_s$ while limiting the solid volume fraction to values less than 0.629 to prevent divergence. Note that $p_c$ is the critical state frictional pressure and many studies assumed the critical state frictional pressure is equal to solid frictional pressure $p_{fr}$, although clearly it is not an accurate assumption.

Up to this point, all of the discussed approaches were based on the von Mises/Mohr–Coulomb law that does not provide any information on how the granular material deforms and flows. Therefore, the von Mises/Coulomb yield criterion cannot model the effect of compressibility phenomena (i.e., changes of bulk density) occurring in the leg pipe and loop seals of the returning system of circulating fluidized bed units and subsequently results in severe underestimation of the exerted frictional viscous forces.

The interparticle contact will be dominant with the increase of volume fraction of particles. For the contact between two spheres the strain changes simultaneously as the particle is deformed, leads to the generation of contact force. The fluctuation velocity of particles $\vartheta$ is decomposed into a translational velocity fluctuation $\boldsymbol{C}$ and a fluctuation velocity of deformation $\boldsymbol{C}_d$. The translational fluctuation kinetic energy is $m < \mathbf{C}^2 > /2$, and the deformation fluctuation kinetic energy is $m < \mathbf{C}_d^2 > /2$. The kinetic energy of motion of particles $m < \vartheta^2 > /2$ is the sum of the translational kinetic energy and the kinetic energy of deformation (Liyan et al. 2015; Huilin 2017).

$$\frac{1}{2}m < \vartheta^2 >= \frac{1}{2}m < \mathbf{C}^2 > + \frac{1}{2}m < \mathbf{C}_d^2 > \tag{2.111}$$

The first part on the right-hand side of Eq. (2.111) represents the translational kinetic energy contribution by translational velocity fluctuations arising from the streaming of particles moving across imaginary shear layer in the flow and direct collisions, and the second part represents the kinetic energy of deformation contribution by displacement fluctuation due to the contact interactions. The particle fluctuation kinetic energy is defined by

$$\frac{1}{2}m < \vartheta^2 >= \frac{3}{2}m\theta_e \tag{2.112}$$

where $\theta_e$ is the generalized granular temperature of particles, and it has the unit of $(\mathrm{m/s})^2$. $< \mathbf{C}^2 > /3$ represents the translational granular temperature, and $< \mathbf{C}_d^2 > /3$ is the configurational temperature. From Eqs. (2.111) and (2.112), it combines the kinetic and configurational contributions, and can serve as a generalized granular temperature to be valid for all granular phase.

$$\theta_e = \theta + \theta_c \tag{2.113}$$

The definition used here is more convenient, since the generalized granular temperature is simply the sum of the variance of the translational fluctuation velocity and fluctuation velocity of deformation of particles. The following correlation is used for describing the relation between the configurational temperature and translational granular temperature as a function of solids volume fractions $\varepsilon_s$, spring coefficient $k_n$ and damping coefficient $\eta_n$.

$$\frac{\theta}{\theta_c} = \frac{10^{(-5.2353+55.3782\alpha_s-357.9726\alpha_s^2+1176.8378\alpha_s^3-1854.8365\alpha_s^4+1126.0306\alpha_s^5)}}{1-(\alpha_s/\alpha_{\max})^{1/3}}$$
$$[1.8657exp(-7.35\times 10^{-4}k_n)][1.024+0.769\eta_n-9.603\eta_n^2] \quad (2.114)$$

The correlation shows the ratio of configurational temperature to translational granular temperature increases monotonously with the increase of solids volume fractions, and decreases with the increases of normal spring coefficient and friction coefficient of particles.

Since the configurational temperature is produced by contact interactions of particles, we anticipate a relationship between the particle pressure and the configurational temperature. Substituting generalized granular temperature for translational granular temperature in expression of particle pressure proposed by Gidaspow (1994), the particle pressure is predicted by

$$p_s = \alpha_s\rho_s\theta + 2\alpha_s^2\rho_s g_0(1+e)(\theta+\theta_c) \quad (2.115)$$

Figure 2.2 shows the measured and simulated particle pressures scaled by the solid particle density $\rho_s$ as a function of solids volume fractions. Campbell and Wang (1991) measured particle pressures using a differential pressure transducer on the wall of a gas-particle fluidized bed. At the high solids volume fractions, the calculated particle pressures based on generalized granular temperature agree with the experimental data. At the low solids volume fractions, the calculated particle pressure levels off and decreases with decreasing solids volume fractions because of the low translational granular temperature in the spout regime of the flat-base spout bed with a jet.

Using Eq. (2.100) and replacing generalized granular temperature for translational granular temperature, the solid viscosity is predicted by

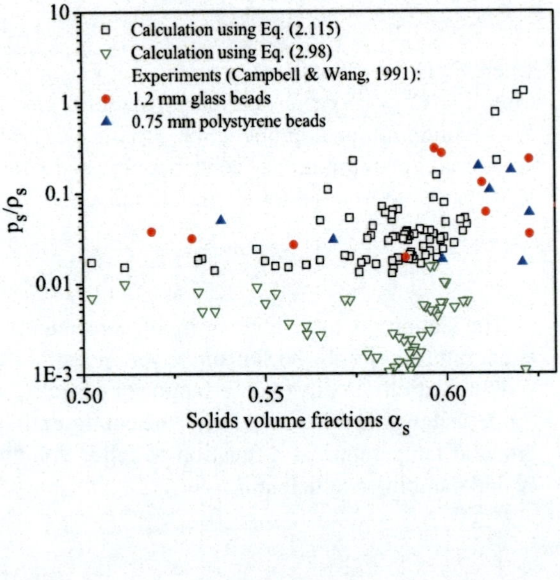


**Fig. 2.2** Profile of measured and simulated particle pressure scales with the density

$$\mu_s = \frac{10\alpha_s d_p \rho_s \sqrt{\pi\theta}}{96(1+e)g_0}[1 + \frac{4}{5}(1+e)\alpha_s g_0]^2 + \frac{4}{5}\alpha_s^2 \rho_s d_p g_0 (1+e)\sqrt{\frac{(\theta+\theta_c)}{\pi}} \tag{2.116}$$

Substituting generalized granular temperature $\theta_e$ for translational granular temperature $\theta$ in expression of kinetic energy dissipation proposed by Gidaspow (1994), the rate of energy dissipation $\gamma$ is predicted by

$$\gamma = \frac{12(1-e^2)g_0}{d_p\sqrt{\pi}}\rho_s\alpha_s^2(\theta+\theta_c)^{3/2} \tag{2.117}$$

The kinetic energy dissipation by deformation is proportional to the square of solid volume fraction and 3/2 power of the configurational temperature. Thus, the kinetic energy dissipation is increased at high solids volume fractions. The sliding and rubbing contact interactions between particles cause the deformation of particles and results in energy dissipations. Hence the energy reflected in the configurational temperature is continually being dissipated away to heat and must be re-supplied from inlet gas phase at the bottom of the bed. As the overlaps derive from the contact interaction of particles, the configurational temperature is created by deformation. Thus, the driving forces from fluid are converted into the kinetic energy of bed material, which is converted to configurational temperature through contact deformation of particles. In the kinetic theory, only the kinetic and collisional contributions to the energy dissipation are accounted for. All the collisions are assumed to be binary and quasi-instantaneous. In regions with high particle volume fractions, multi-particle contact interactions dominate the energy dissipation mechanism, and much more energy will be dissipated. Unfortunately, the contribution of energy dissipation by the deformation of contact interaction does not take into account the calculation of energy dissipation in the original kinetic theory of granular flow.

Substituting generalized granular temperature $\theta_e$ for translational granular temperature $\theta$ in Eq. (2.104), the granular conductivity of fluctuating energy is predicted by

$$k_s = \frac{150\rho_s d_p\sqrt{\pi\theta}}{384(1+e)g_0}[1 + \frac{6}{5}\alpha_s g_0(1+e)]^2 + 2\rho_s\alpha_s^2 d(1+e)g_0\sqrt{\frac{(\theta+\theta_c)}{\pi}} \tag{2.118}$$

Note that the contribution of the fluctuating energy of deformation does not take into account the calculation of granular conductivity in the kinetic-frictional stress models used in the numerical simulations of gas–solid flow (e.g. Huilin et al. 2004; ANSYS 2011).

## 2.7 Boundary Conditions

The dissipation of solids turbulent kinetic energy by collisions with the wall is specified by the particle–wall restitution coefficient, $e_{sw}$. A high value of specularity coefficient, $\varphi$, implies high production at the wall and a value of $e_{sw}$ close to unity implies low dissipation of granular energy at the wall. It is expected that the specularity coefficient and the particle–wall restitution coefficient need to be calibrated for a given gas/particle flow system because the specularity coefficient cannot be measured and $e_{sw}$ can be measured only with some difficulty (Johnson and Jackson 1987).

$$v_{s,w} = -\frac{6\mu_s \alpha_{s,\max}}{\sqrt{3}\pi\varphi\rho_s\alpha_s g_0\sqrt{\theta}}\frac{\partial v_{s,w}}{\partial x} \tag{2.119}$$

$$\theta_w = -\frac{4\alpha_{s,\max}k_s\theta}{\sqrt{3}\pi\left(1 - e_{sw}^2\right)\alpha_s\rho_s g_0\theta^{3/2}}\frac{\partial\theta_w}{\partial x} + \frac{4\varphi v_{s,slip}^2}{6\left(1 - e_{sw}^2\right)} \tag{2.120}$$

## 2.8 Kinetic Theory of Particles Mixture

Gas-particle fluidize beds are composed of particles of different properties, in which segregation by size or density may occur during the fluidization. The coal particles compose of small solid particles, normally with diameters of less than (8.0–12.0) mm in the case of a power plant (Huilin 2017), although they can be larger in industrial fluidized bed applications. The remainder is the inert bed material. The large particles held at the bottom of the combustion chamber. The small particles are fluidized at the dilute zone in the upper part of the chamber. It has important implications in the mixing/segregation behavior of coal particles in the fluidized bed combustors.

The extension of the kinetic theory was proposed by Jenkins and Mancini (1989) for a binary mixture of smooth nearly elastic spheres assuming a mixture of particles having different mass and size. They assumed that the two species of particles in the mixture have an average granular temperature. Farrell et al. (1986) speculated that the assumption of equal granular temperature may not be true. Experiments with a binary mixture consisting of a 170 μm steel particles and 450 μm glass beads were done by Gidaspow and Huilin (1997) in a liquid–solid fluidized bed. Figure 2.3 shows the granular temperatures of steel and glass in the mixture and those for steel and glass alone. The experimental data clearly show that the equipartition of kinetic energy does not hold. The granular temperatures of the steel particles and the glass beads in the mixture are not equal. The granular temperature of steel particles in the mixture is larger than that of steel particles alone. The granular temperature of the glass particles in the mixture is only slightly less than that of the glass beads alone. Hence a separate balance of granular temperature is needed.

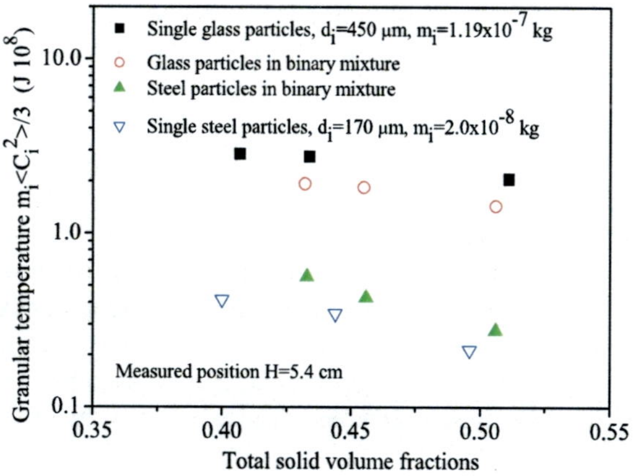


**Fig. 2.3** Test of equipartition of oscillating kinetic energy

A kinetic theory model of binary mixture for two-size particles with different granular temperatures was developed (Huilin et al. 2000; 2001). Each particle type is represented by a phase, with an average velocity and a fluctuating energy or granular temperature. This means that the interaction between the different particle spices is at the interface. Garzo and Dufty (2002) solved the kinetic equation for systems away from equilibrium. This approach could capture not only the energy non-equipartition but also the flow behavior for a wide range of restitution coefficients. Such a model is restricted to dilute systems where the radial distribution function is close to unity. Iddir and Arastoopour (2005) extended the kinetic theory to a multi-type (size and/or density) mixture, assuming a non-Maxwellian velocity distribution and energy non-equipartition. They solved the Boltzmann's equation for each particulate phase using the Chapman-Enskog procedure by adding a perturbation to the Maxwellian velocity distribution function. Garzo et al. (2007) proposed the constitutive relations based on the kinetic theory with unequal component granular temperatures of multi-component mixture. The derived transport coefficients were correlated with component granular temperatures which were used in the dilute and dense flows.

### *2.8.1 Boltzmann Equations for PhaseI in a Mixture of M Phases*

Boltzmann equations can be derived for each component or each phase. Considering binary collisions only, the equations for the frequency distributions $f_i$ are ($i = 1, 2, K, M$)

$$\frac{\partial f_i}{\partial t} + c_i \cdot \frac{\partial f_i}{\partial r} + F_i \cdot \frac{\partial f_i}{\partial c_i} = \sum_{k=1}^{M} \iint \left(f_i' f_k' - f_i f_k\right) c_{ik} \cdot \mathbf{k}\, d_{ik}^2\, d\mathbf{k} d\mathbf{c}_k \qquad (2.121)$$

Let the granular temperature for phase $i$, $\theta_i$ be defined as the average variance of velocity distribution of phase $i$ over the 3 directions.

$$\theta_i = \frac{1}{3} m_i < \mathbf{C}_i^2 > \tag{2.122}$$

where $\mathbf{C}_i$ is the fluctuating velocity for spheres $i$ relative to the hydrodynamic velocity. In contrast with the one thermal temperature for gases (Chapman and Cowling 1961) and one granular temperature for binary granular mixture (Farrell et al. 1986; Jenkins and Mancini 1989), we assumed that each species has a different granular temperature. Then the Maxwellian distribution becomes essentially that used in gas theory with the Boltzmann constant equal to unity, but with unequal granular temperatures.

$$f_i^{(0)} = n_i \left( \frac{m_i}{2\pi\theta_i} \right) \exp\left[ -\frac{m_i}{2\theta_i} (c_i - v_i) \cdot (c_i - v_i) \right] \tag{2.123}$$

The number of binary collisions between particles 1 and 2 is given by

$$N_{ik} = \iiint f_{ik}^{(2)}(\mathbf{r}_{1i}, \mathbf{c}_{1i}, \mathbf{r}_{2k}, \mathbf{c}_{2k})(\mathbf{c}_{21ki} \cdot \mathbf{k} d_{ik}^2) d\mathbf{k} d\mathbf{c}_{1i} d\mathbf{c}_{2k} \tag{2.124}$$

where the distribution is defined such that $f_{ik}^{(2)} d\mathbf{c}_{1i} d\mathbf{r}_1 d\mathbf{c}_{2k} d\mathbf{r}_{2k}$ is the probability for finding a pair of particles 1 and 2 in the volume $d\mathbf{r}_{1i} d\mathbf{r}_{2k}$ with velocities $\mathbf{c}_{1i}$ and $\mathbf{c}_{2k}$, $\mathbf{k}$ is the unit vector directed from the center of particle 2 to the center of particle1 at collision, $d_{ik}$ is the mean particle diameter of particle 1 and 2, $\mathbf{c}_{21}$ is the relative velocity between particle 1 and particle 2. With the assumption of chaos, the pair distribution is related to the single particle velocity distribution functions as

$$f_{ik}^{(2)} = f_{ik}^{(2)}(\mathbf{r}_{1i}, \mathbf{c}_{1i}, \mathbf{r}_{2k}, \mathbf{c}_{2k}) = f_i(\mathbf{r}, \mathbf{c}_{1i}) f_k(\mathbf{r}, \mathbf{c}_{2k}) \tag{2.125}$$

The number of binary collisions can then be written as:

$$N_{ik} = \frac{n_i n_k}{(2\pi)^3} \left( \frac{m_i m_k}{\theta_i \theta_k} \right)^{3/2} \iiint (\mathbf{c}_{21ki} \cdot \mathbf{k} d_{ik}^2) \exp[-\frac{m_i \mathbf{c}_i^2}{2\theta_i} - \frac{m_k \mathbf{c}_{2k}^2}{2\theta_k}] d\mathbf{k} d\mathbf{c}_1 d\mathbf{c}_2 \tag{2.126}$$

Transforming to the relative velocity $\mathbf{c}_{21,ik}$ and to the mass-center velocity, $\mathbf{G}$, using the collisional relations in a similar manner as done by Chapman and Cowling (1961), and expanding it in a Taylor series, the integral can be written as

$$N_{ik} = \frac{\pi d_{ik}^2}{(2\pi)^3} \left( \frac{m_i m_k}{\theta_i \theta_k} \right)^{3/2} \iint \mathbf{c}_{21ki} \exp[-(A\mathbf{G}_{ki}^2 + 2B\mathbf{G}_{ki}\mathbf{c}_{21ki} + D\mathbf{c}_{21ki}^2)] d\mathbf{G}_{ki} d\mathbf{c}_{21ki} \tag{2.127}$$

where the coefficients A, B and D are

$$A = \frac{m_i\theta_k + m_k\theta_i}{2\theta_i\theta_k}, \quad B = \frac{m_i m_k(\theta_i - \theta_k)}{2m_o\theta_i\theta_k}, \quad D = \frac{m_i m_k(m_i\theta_i - m_k\theta_k)}{2m_0^2\theta_i\theta_k} \tag{2.128}$$

where $m_o$ is the sum of $m_a$ and $m_b$. Carrying out the integration by parts, the result is:

$$N_{ik} = \frac{\sqrt{\pi}}{4} d_{ik}^2 n_i n_k \left(\frac{m_i m_k}{\theta_i\theta_k}\right)^{3/2} \frac{1}{A^{3/2}D^2}\left(1 - 3\frac{B}{\sqrt{AC}} + 6\frac{B^2}{AD} - 10\frac{B^3}{AC^{3/2}} + 15\frac{B^4}{(AD)^2}\cdots\right) \tag{2.129}$$

## 2.8.2 Dense Transport Theorem

The dense transport theorem is obtained as in Chapman and Cowling (1961) for the property $\psi$ ($i = i$ or $j$).

$$\frac{\partial(n_i\psi_i)}{\partial t} + \nabla\cdot(n_i < \mathbf{c}_i\psi_i > + \sum_{k=i,j}\mathbf{P}_{cik}) = n_i < \mathbf{F}_i\frac{\partial\psi_i}{\partial\mathbf{c}_i} > + \sum_{k=i,j} N_{cik} \tag{2.130}$$

The collisional pressure or energy contribution is given by

$$P_{cik} = -\frac{d_{ik}^3}{2}\iiint\limits_{k\cdot c_{12,ik}>0}(\psi'_{1i} - \psi_{1i})(\mathbf{k}\cdot\mathbf{c}_{12ik})\mathbf{k}f_{ik}^{(2)}(\mathbf{r} - \frac{1}{2}d_{ik}\mathbf{k}, \mathbf{c}_{1i}, \mathbf{r} + \frac{1}{2}d_{ik}\mathbf{k}, \mathbf{c}_{2k})d\mathbf{k}d\mathbf{c}_{1i}d\mathbf{c}_{2k} \tag{2.131a}$$

The source like contribution is

$$N_{cik} = \frac{d_{ik}^2}{2}\iiint\limits_{\mathbf{k}\cdot\mathbf{c}_{12ik}>0}(\psi'_{1i} + \psi'_{2k} - \psi_{1i} - \psi_{2k})(\mathbf{k}\cdot\mathbf{c}_{12ik})\mathbf{k}f_{ik}^{(2)}(\mathbf{r} - \frac{1}{2}d_{ik}\mathbf{k}, \mathbf{c}_{1i}, \mathbf{r} + \frac{1}{2}d_{ik}\mathbf{k}, \mathbf{c}_{2k})d\mathbf{k}d\mathbf{c}_{1i}d\mathbf{c}_{2k} \tag{2.131b}$$

where the binary collisional frequency distribution $f^{(2)}$ is evaluated as in Eq. (2.125) with the particle diameter replaced by the mean particle diameter $d_{p_{ik}}$.

The contribution due to the collisions between unlike particles is

$$N_{c_{ik}} = \frac{d_{p_{ik}}^2}{2}\int\limits_{\mathbf{k}\cdot\mathbf{c}_{12}>0}\left(\psi'_{1_i} - \psi_{1_i}\right)(\mathbf{k}\cdot\mathbf{c}_{12ik})f_{ik}^{(2)}(\mathbf{r} - \frac{1}{2}d_{ik}\mathbf{k}, \mathbf{c}_{1i}, \mathbf{r} + \frac{1}{2}d_{ik}\mathbf{k}, \mathbf{c}_{2k})d\mathbf{k}d\mathbf{c}_{1i}d\mathbf{c}_{2k}$$

$$+\frac{d_{pik}^2}{2}\int_{\mathbf{k}\cdot\mathbf{c}_{12}>0}\left(\psi'_{2_i}-\psi_{2_i}\right)(\mathbf{k}\cdot\mathbf{c}_{12ik})\mathbf{k}f_{ik}^{(2)}(\mathbf{r}-\frac{1}{2}d_{ik}\mathbf{k},\mathbf{c}_{1i},\mathbf{r}+\frac{1}{2}d_{ik}\mathbf{k},\mathbf{c}_{2k})d\mathbf{k}d\mathbf{c}_{1i}d\mathbf{c}_{2k} \tag{2.132}$$

## 2.8.3 Conservation Equations

The balance laws in the mixture are obtained as follows. Let $\psi_i = m_i$ to obtain the mass balance for phase $i$ with no phase change. Then the dense transport theorem gives:

$$\frac{\partial}{\partial t}(n_i m_i) + \nabla\cdot(n_i m_i \mathbf{v}_{si}) = 0 \tag{2.133}$$

Let $\psi_i = m_i\mathbf{c}_j$ to obtain the balance of linear momentum for phase $i$:

$$\frac{\partial}{\partial t}(\alpha_{si}\rho_{si}\mathbf{v}_{si}) + \nabla\cdot(\alpha_{si}\rho_{si}\mathbf{v}_{si}\mathbf{v}_{si}) = -\nabla\cdot(\mathbf{P}_{ki}+\mathbf{P}_{ci}) + \alpha_{si}\rho_{si}\mathbf{F}_i + N_{ci}(m_i\mathbf{c}_i) \tag{2.134}$$

Here $P_k$ and $P_c$ are the pressure tensor of the transport and collisional contributions. Let $\psi_i = \mathbf{c}_i^2/2$ and $<\mathbf{C}_i^2>= 3\theta_i$ to obtain the balance of fluctuation energy for phase $i$ as follows:

$$\frac{3}{2}[\frac{\partial}{\partial t}(n_i\theta_i) + \nabla\cdot(n_i\mathbf{v}_{si}\theta_i)] = (\mathbf{P}_{ki}+\mathbf{P}_{ci}) : \nabla\mathbf{v}_{si} - \nabla(\mathbf{q}_{ki}+\mathbf{q}_{ci}) + N_{ci}(\frac{1}{2}m_i\mathbf{c}_i^2) + \alpha_{si}\rho_{si} < \mathbf{F}_i\cdot\mathbf{C}_i > \tag{2.135}$$

## 2.8.4 Constitutive Equations

In order to calculate the collisional terms appearing in the balance laws, definite forms of the pair distributions at contact are needed. If it is assumed that collisions between two particles are only slightly influenced by the presence of other particles, the pair distribution functions can be expressed in terms of the product of two single particle velocity distribution functions:

$$f_{ik}^{(2)}(\mathbf{r}-\frac{1}{2}d_{ik}\mathbf{k},\mathbf{c}_{1i},\mathbf{r}+\frac{1}{2}d_{ik}\mathbf{k},\mathbf{c}_{2k}) = g_{ik}(d_i,d_k)f_i(\mathbf{r}-\frac{1}{2}d_{ik}\mathbf{k},\mathbf{c}_{1i})f_k(\mathbf{r}+\frac{1}{2}d_{ik}\mathbf{k},\mathbf{c}_{2k}) \tag{2.136}$$

The factor $g_{ik}$ is the equilibrium radial distribution function of two spheres, one of species i and the other of species k, at contact. The expression for the radial distribution at contact for mixtures of hard spheres that is in best agreement with numerical simulations is that of Mansoori et al. (1971). For binary mixture it can be written as

$$g_{ik} = \frac{1}{1-\alpha_{sik}} + \frac{3d_i d_k}{d_i + d_k}\frac{\xi}{(1-\alpha_{sik})^2} + 2\left(\frac{d_i d_k}{d_i + d_k}\right)^2 \frac{\xi^2}{(1-\alpha_{sik})^3} \tag{2.137}$$

where $\xi = 4\pi\left(n_i d_i^2 + n_k d_k^2\right)/3$ and $\alpha_{sik}$ is the total volume fraction. Expanding the distribution function in Taylor series, it gives

$$f_{ik}^{(2)}(\mathbf{r} - \frac{1}{2}d_{ik}\mathbf{k}, \mathbf{c}_{1i}, \mathbf{r} + \frac{1}{2}d_{ik}\mathbf{k}, \mathbf{c}_{2k}) = g_{ik}(d_i, d_k)[f_i(\mathbf{r}, \mathbf{c}_{1i}) f_k(\mathbf{r}, \mathbf{c}_{2k})$$
$$+ \frac{1}{2}(d_{ik}\cdot\mathbf{k}) f_i(\mathbf{r}, \mathbf{c}_{1i}) f_k(\mathbf{r}, \mathbf{c}_{2k}) \nabla \ln \frac{f_k(\mathbf{r}, \mathbf{c}_{2k})}{f_i(\mathbf{r}, \mathbf{c}_{1i})}] \tag{2.138}$$

For a Maxwellian distribution the kinetic part of the stress tensor for species $i$ was defined as:

$$\mathbf{P}_{ki}^{(0)} = n_i m_i < \mathbf{C}_i^{(0)}\mathbf{C}_i^{(0)} >= \int n_i m_i \mathbf{C}_i \mathbf{C}_i f_i^{(0)} d\mathbf{c}_i \tag{2.139}$$

Carrying out the integration, the first approximations for the stress tensor gives the equation of state, that is

$$\mathbf{P}_{ki}^{(0)} = \mathrm{P}_i \mathbf{I} = n_i \theta_i \mathbf{I} \tag{2.140}$$

where $\boldsymbol{I}$ is a unit tensor. From the transport equation for kinetic energy, the first approximation to the heat flux for species $i$ is given by

$$\mathbf{q}_{ki}^{(0)} = \int \frac{1}{2} n_i m_i \mathbf{C}_i^2 \mathbf{C}_i f_i^{(0)} d\mathbf{c}_i \tag{2.141}$$

But note that $< \mathbf{C}_i^k >= 0$ for every odd integer $k$. Thus the first approximation to the heat flux gives

$$\mathbf{q}_{ki}^{(0)} = 0 \tag{2.142}$$

Substitution of Eq. (2.138) into (2.131) gives an expression for the collisional stress tensor shown below:

$$\mathbf{P}_{ci} = \sum_k (\mathbf{P}_{cik}^1 + \mathbf{P}_{cik}^2)$$

$$= \sum_k [-\frac{1}{2} g_{ik} d_{ik}^3 \iiint_{\mathbf{c}_{12ik}\cdot\mathbf{k}>0} (\psi'_{1i} - \psi_{1i}) f_i f_k \mathbf{k}(\mathbf{k}\cdot\mathbf{c}_{12ik}) d\mathbf{k} d\mathbf{c}_{1i} d\mathbf{c}_{2k}$$

$$- \frac{1}{4} g_{ik} d_{ik}^4 \iiint_{c_{12ik}\cdot k>0} (\psi'_{1i} - \psi_{1i}) f_i f_k \nabla \ln \frac{f_k}{f_i} \mathbf{k}\cdot(\mathbf{k}\cdot\mathbf{c}_{12ik}) d\mathbf{k} d\mathbf{c}_{1i} d\mathbf{c}_{2k}] \quad (2.143)$$

The collisional dynamics showed that

$$\mathbf{c}'_{1i} - \mathbf{c}_{1k} = -\frac{m_k}{m_0}(1+e)(\mathbf{k}\cdot\mathbf{c}_{12ik})\mathbf{k} \quad (2.144)$$

Letting $\psi_{1i} = m_i \mathbf{C}_{1i}$, the first term in Eq. (2.143) becomes:

$$\mathbf{P}_{cik}^1 = \frac{\pi}{15} g_{ik} d_{ik}^3 \frac{m_i m_k}{m_0}(1+e_{ik}) \iint_{\mathbf{c}_{12ik}\cdot\mathbf{k}>0} f_i f_k (2\mathbf{c}_{12ik}\mathbf{c}_{12ik} + \mathbf{c}_{12ik}^2 \mathbf{I}) d\mathbf{c}_{1i} d\mathbf{c}_{2k} \quad (2.145)$$

where $e_{ik}$ is the restitution coefficient between species $i$ and $k$, $f_i$ and $f_k$ are the velocity distribution of species $i$ and $k$ which are assumed to be Maxwellian. Evaluating this integral using the same method as for the number of collisions, the result becomes

$$\mathbf{P}_{c,ik}^1 = \frac{\pi}{48}(1+e_{ik}) d_{ik}^3 g_{ik} \left(\frac{m_i m_k}{\theta_i \theta_k}\right)^{3/2} \frac{n_i m_i n_k m_k}{m_o A^{3/2} D^{5/2}} \left(1 + 6\frac{B^2}{AD} + 15\frac{B^4}{(AD)^2} \cdots\right)\mathbf{I} \quad (2.146)$$

The total pressure for species $i$ is given by:

$$P_i = n_i \theta_i + \sum_k \mathbf{P}_{cik}^1 \quad (2.147)$$

and the solid pressure, or equation of state for mixture, is of course given by:

$$P_s = \sum_i (n_i \theta_i + \sum_k \mathbf{P}_{cik}^1) = \sum_i n_i \theta_i + \sum_i \sum_k \mathbf{P}_{cik}^1 \quad (2.148)$$

The second term in the Eq. (2.143), $\mathbf{P}_{cik}^2$, is integrated over $\mathbf{k}$ by using the second integration identity provided by Chapman and Cowling (1961). This gives:

$$\mathbf{P}_{cik}^2 = -\mathbf{P}_{cik}^1 \frac{d_{ik}}{3}\left(\frac{2m_i m_k(\theta_i+\theta_k)^2}{\pi\theta_i\theta_k(m_i\theta_i + m_k\theta_k)}\right)^{1/2}\left(\frac{6}{5}\nabla^s \mathbf{v}_s + \nabla\cdot\mathbf{v}_s \mathbf{I}\right) \quad (2.149)$$

where $\nabla^s$ means the rate of shear tensor. From the Eq. (2.149), the viscosity of phase $i$ in the $m$ phases is as follows

$$\mu_{ci} = \sum_{k=1}^{m} \mu_{cik} \tag{2.150}$$

$$\mu_{cik} = \frac{2\pi d_i^4 n_i n_k (1+e_{ik}) g_{ik}}{15} \sqrt{\frac{m_i m_k \theta_i \theta_k (\theta_i + \theta_k)^2}{2\pi (m_i\theta_i + m_k\theta_k)^3}} \left[ \frac{m_o^2 \theta_i \theta_k}{(m_i\theta_i + m_k\theta_k)(m_i\theta_k + m_k\theta_i)} \right]^{3/2} \left( 1 + 6\frac{B^2}{AD} + 15\frac{B^4}{(AD)^2} \ldots \right) \tag{2.151}$$

The mixture viscosity is the sum of the phase viscosities

$$\mu_c = \sum_{i=1}^{M} \sum_{k=1}^{M} \mu_{c,ik} \tag{2.152}$$

The kinetic viscosity for dilute flow is

$$\mu_{i,kin} = \frac{5}{16 d_i^2} \left( \frac{m_i^2}{\pi} \theta_i \right)^{1/2} \tag{2.153}$$

The momentum source contribution in Eq. (2.132) can be written

$$N_{ik}(m_i c_{1i}) = \phi_{ik}^1 + \phi_{ik}^2 \tag{2.154}$$

where

$$\phi_{ik}^1 = -\frac{m_i m_k}{m_o} d_{ik}^2 (1+e_{ik}) \int\limits_{\mathbf{k}\cdot\mathbf{c}_{12}>0} (\mathbf{k}\cdot\mathbf{c}_{12,ik})^2 \mathbf{k} f_i f_k d\mathbf{k} d\mathbf{c}_{1i} d\mathbf{c}_{2k} \tag{2.155a}$$

$$\phi_{ik}^2 = -\frac{m_i m_k}{m_o} \frac{d_{ik}^3}{2} (1+e_{ik}) \int\limits_{\mathbf{k}\cdot\mathbf{c}_{12}>0} (\mathbf{k}\cdot\mathbf{c}_{12ik})^2 \mathbf{k} f_i f_k \mathbf{k} \nabla \ln \frac{f_k}{f_i} d\mathbf{k} d\mathbf{c}_{1i} d\mathbf{c}_{2k} \tag{2.155b}$$

Following directly the same procedure as used by Chapman and Cowling (1961), the final results are:

$$\phi_{ik}^1 = \mathbf{P}_{cik}^1 \left[ \frac{3}{d_{ik}} \left( \frac{2 m_i m_k (m_i\theta_i + m_k\theta_k)}{\pi m_o^2 \theta_i \theta_k} \right)^{1/2} (\mathbf{v}_{sk} - \mathbf{v}_{si}) \right] \tag{2.156a}$$

$$\begin{aligned} \phi_{ik}^2 = \mathbf{P}_{cik}^1 \Bigg\{ & \nabla \ln \frac{n_i}{n_k} + 3\nabla \left( \frac{\ln \theta_k}{\ln \theta_i} \right) + \frac{\theta_i \theta_k}{(m_i\theta_k + m_k\theta_k)} \left( \frac{m_i \nabla \ln \theta_i}{\theta_i} - \frac{m_k \nabla \ln \theta_k}{\theta_k} \right) \\ & + \frac{5}{3} \frac{\theta_i \theta_k}{(m_i\theta_i + m_k\theta_k)} \left( \frac{m_k \nabla \ln \theta_i}{\theta_i} - \frac{m_i \nabla \ln \theta_k}{\theta_k} \right) \Bigg\} \end{aligned} \tag{2.156b}$$

The energy dissipation for species $i$ is given by:

$$\begin{aligned}\gamma_i &= N_{ci}\left(\frac{1}{2}m_i\mathbf{C}_i^2\right) = \sum_k N_{cik}\left(\frac{1}{2}m_i\mathbf{c}_{1i}^2\right) \\ &= \sum_k \frac{1}{2}d_{ik}^2 \iiint\limits_{\mathbf{c}_{12}\cdot\mathbf{k}>0} (\psi'_{1i} + \psi'_{2k} - \psi_{1i} - \psi_{2k})(\mathbf{k}\cdot\mathbf{c}_{12ik}) f_{ik}^{(2)} d\mathbf{k} d\mathbf{c}_{1i} d\mathbf{c}_{2k} = \\ &= \sum_k (\gamma_i^1 + \gamma_i^2)\end{aligned} \tag{2.157}$$

where

$$\gamma_i^1 = \frac{d_{ik}^2}{2}\frac{m_i m_k}{m_o}(e_{ik}^2 - 1)g_{ik} \iiint\limits_{\mathbf{c}_{12}\cdot\mathbf{k}>0} (\mathbf{k}\cdot\mathbf{c}_{12ik})^3 f_i f_k d\mathbf{k} d\mathbf{c}_{1i} d\mathbf{c}_{2k} \tag{2.158a}$$

$$\gamma_i^2 = \frac{d_{ik}^2}{8}\frac{m_i m_k}{m_o}(e_{ik}^2 - 1)g_{ik} \iiint\limits_{\mathbf{c}_{12}\cdot\mathbf{k}>0} (\mathbf{k}\cdot\mathbf{c}_{12ik})^3 \mathbf{k} f_i f_k \nabla \ln\frac{f_k}{f_i} d\mathbf{k} d\mathbf{c}_{1i} d\mathbf{c}_{2k} \tag{2.158b}$$

These integral can be evaluated using the integral theorems and the same technique as used by Gidaspow (1994):

$$\gamma_i^1 = \frac{\sqrt{\pi}}{16} d_{ik}^2 (e_{ik}^2 - 1) g_{ik} \frac{m_i m_k}{m_o}\left(\frac{m_i m_k}{\theta_i \theta_k}\right)^{3/2} \frac{n_i n_k}{A^{3/2} D^3}\left(1 + 6\frac{B^2}{AD} + 15\frac{B^4}{(AD)^2}\cdots\right) \tag{2.159a}$$

$$\gamma_i^2 = \frac{\pi}{4}(e_{ik} - 1)\mathbf{P}_{cik}\frac{m_o(\theta_i + \theta_k)}{(m_i\theta_i + m_k\theta_k)}\nabla\cdot\mathbf{v}_i \tag{2.159b}$$

The collisional heat flux for species $i$ is given by:

$$\begin{aligned}\mathbf{q}_{ci} &= \sum_k P_{cik}\left(\frac{1}{2}m_i\mathbf{c}_{ic}^3\right) \\ &= \frac{1}{2}g_{ik}d_{ik}^3 \iiint\limits_{\mathbf{c}_{12}\cdot\mathbf{k}>0} (\psi'_{1i} - \psi_{1i}) f_i f_k \mathbf{k}(\mathbf{c}_{12ik}\cdot\mathbf{k}) d\mathbf{k} d\mathbf{c}_{1i} d\mathbf{c}_{2k} \\ &+ \frac{1}{4}g_{ik}d_{ik}^4 \iiint\limits_{\mathbf{c}_{12}\cdot\mathbf{k}>0} (\psi'_{1i} - \psi_{1i}) f_i f_k \mathbf{k}\nabla \ln\frac{f_k}{f_i}\mathbf{k}(\mathbf{c}_{12ik}\cdot\mathbf{k}) d\mathbf{k} d\mathbf{c}_{1i} d\mathbf{c}_{2k} = \sum_k (\mathbf{q}_{cik}^1 + \mathbf{q}_{cik}^2)\end{aligned} \tag{2.160}$$

Following the same procedure as used for calculating the energy dissipation (Chapman and Cowling 1961; Gidaspow 1994), the final results are:

$$\mathbf{q}_{cik}^1 = -\frac{9}{5}\mathbf{P}_{cik}(1 + e_{ik})\frac{m_k}{m_o}(\mathbf{v}_k - \mathbf{v}_i) \tag{2.161a}$$

$$\mathbf{q}_{cik}^2 = -\mathbf{P}_{cik} d_{ik}(1+e_{ik})\left\{\left(\frac{m_k\theta_i\theta_k}{2\pi m_i(m_i\theta_i+m_k\theta_k)}\right)^{1/2}\left[\nabla \ln\frac{n_i}{n_k}+3\nabla\left(\frac{\ln\theta_k}{\ln\theta_i}\right)\right]\right.$$
$$+3\left(\frac{\theta_i\theta_k}{2\pi m_i m_k(m_i\theta_i+m_k\theta_k)}\right)^{1/2}\left(\frac{m_k\theta_i\theta_k}{m_i\theta_k+m_k\theta_i}\right)\left(\frac{m_i\nabla\ln\theta_i}{\theta_i}-\frac{m_k\nabla\ln\theta_k}{\theta_k}\right)$$
$$\left.+\frac{5}{\sqrt{2\pi}}\left(\frac{m_k\theta_i\theta_k}{m_i(m_i\theta_i+m_k\theta_k)}\right)^{3/2}\left(\frac{m_k\nabla\ln\theta_i}{\theta_i}-\frac{m_i\nabla\ln\theta_k}{\theta_k}\right)\right\} \quad (2.161b)$$

### 2.8.5 Summary of Model Equations

CFD codes using essentially the theory summarized in Table 3.1 were written to model a bubbling bed with a binary mixture of particles (Huilin et al. 2003) and a riser (Huilin and Gidaspow 2003) with again a binary mixture. In the riser, as expected the larger particles were near the bottom. The code computed a larger variance of velocities in the direction of flow than in the radial direction. The CFD code can be modified to incorporate this new theory. Separated phase granular temperature equations must be added. Then in the separate phase momentum equations the solids stress will be computed from the derivatives of the phase *i* pressures and viscosities. The kinetic theory approach was extended to flow of particles rotation and anisotropic flow using the method of moments. In addition, the kinetic theory has been extended to cohesive particle flow (Kim and Arastoopour 2002; Huilin et al. 2010).

Table 3.1 A summary of governing equations and constitutive equations of particles mixture

A. Governing equations:

(a) Conservation of mass for gas phase

$$\frac{\partial}{\partial t}(\alpha_g\rho_g)+\nabla\cdot(\alpha_g\rho_g\mathbf{v}_g)=0 \quad \text{(T1)}$$

(b) Conservation of mass for particle component m (m = 1, 2,….M)

$$\frac{\partial}{\partial t}(\alpha_m\rho_m)+\nabla.(\alpha_m\rho_m\mathbf{v}_m)=0 \quad \text{(T2)}$$

(c) Conservation of momentum for gas phase

$$\frac{\partial}{\partial t}(\alpha_g\rho_g\mathbf{v}_g)+\nabla\cdot(\alpha_g\rho_g\mathbf{v}_g\mathbf{v}_g)=-\alpha_g\nabla p+\nabla\cdot\tau_g+\alpha_g\rho_g g+\sum_{m=1}^{M}\beta_{gm}(\mathbf{v}_m-\mathbf{v}_g) \quad \text{(T3)}$$

(d) Conservation of momentum for particle component m

$$\frac{\partial}{\partial t}(\alpha_m \rho_m \mathbf{v}_m) + \nabla \cdot (\alpha_m \rho_m \mathbf{v}_m \mathbf{v}_m) = - \alpha_m \nabla p + \nabla \cdot \tau_m + \nabla p_m + \alpha_m \rho_m g + \beta_{gm}\left(\mathbf{v}_g - \mathbf{v}_m\right) + \sum_{n=1, m\neq n}^{M} \beta_{mn}(\mathbf{v}_n - \mathbf{v}_m) \tag{T4}$$

(e) Conservation of fluctuating energy for particle component m

$$\frac{3}{2}\left[\frac{\partial}{\partial t}(\alpha_m \rho_m \theta_m) + \nabla \cdot (\alpha_m \rho_m \theta_m \mathbf{v}_m)\right] = (\tau_m : \nabla \cdot \mathbf{v}_m) + \nabla \cdot q_m - \gamma_m - 3\beta_{gm}\theta_m \tag{T5}$$

B. Constitutive equations:

(a) Solid stress for component *m*:

$$\tau_m = \alpha_m \mu_m \left[\nabla \mathbf{v}_m + (\nabla \mathbf{v}_m)^T\right] + \alpha_m \left(\lambda_m - \frac{2}{3}\mu_m\right)(\nabla \cdot \mathbf{v}_m)\mathbf{I} \tag{T6}$$

(b) Solid phase pressure for component m:

$$p_m = \alpha_m \rho_m \theta_m + \sum_{n=1}^{M} p_{nm} \tag{T7}$$

$$p_{nm} = \frac{\pi(1+e_{nm})d_{nm}^3 g_{nm} n_n n_m m_n m_m m_o \theta_n \theta_m}{3(m_n^2\theta_n + m_m^2\theta_m)}\left[\frac{m_o^2 \theta_n \theta_m}{(m_n^2\theta_n + m_m^2\theta_m)(\theta_n + \theta_m)}\right]^{3/2}(1 + 6\Delta^2 + 15\Delta^4 \ldots) \tag{T8}$$

$$\Delta = \frac{(m_m\theta_m - m_n\theta_n)}{[(m_m^2\theta_m^2 + m_n^2\theta_n^2) + \theta_m\theta_n(m_m^2 + m_n^2)]^{1/2}} \tag{T9}$$

(c) Shear viscosity of component m

$$\mu_m = \sum_{n=1}^{M} p_{nm}\frac{d_{nm}}{5}\sqrt{\frac{2(m_m\theta_m + m_n\theta_n)^2}{\pi\theta_n\theta_m(m_m^2\theta_m + m_n^2\theta_n)}} + \frac{2\mu_{dil,m}}{\frac{1}{n_n}\sum_{n=1}^{M}(1+e_{nm})g_{nm}}\left[1 + \frac{4}{5}\sum_{n=1}^{M}(1+e_{nm})\alpha_n g_{nm}\right]^2 \tag{T10}$$

$$\mu_{dil,m} = \frac{5\sqrt{\pi}}{96} d_m \rho_m \theta_{av}^{1/2} \tag{T11}$$

$$\theta_{av} = \frac{2\theta_m}{\left\{\sum_{n=1}^{M}\left(\frac{n_n d_n^2}{n_m d_m^2}\right)\left[\frac{m_o^2\theta_n}{(m_m^2\theta_m + m_n^2\theta_n)}\right]^{1/2}\left[\frac{m_o^2\theta_n\theta_m}{(m_m^2\theta_m + m_n^2\theta_n)(\theta_m + \theta_n)}\right]^{3/2}(1 + 6\Delta^2 + 15\Delta^4 + \cdots)\right\}^2} \tag{T12}$$

(d) Bulk viscosity of component m

$$\lambda_m = \sum_{n=1}^{M} \frac{d_{nm}}{3} [\frac{2(m_m\theta_m + m_n\theta_n)^2}{\pi\theta_m\theta_n(m_m^2\theta_m + m_n^2\theta_n)}]^{1/2} p_{nm} \tag{T13}$$

(e) Collisional kinetic energy flux for component $m$

$$\begin{aligned} q_m = \sum_{n=1}^{M} p_{nm}(1+e_{nm})\Big\{ & \frac{9m_n}{5m_o}(\mathbf{v}_n - \mathbf{v}_m) + d_{nm}[(\frac{2m_n^2\theta_n}{\pi(m_m^2\theta_m + m_n^2\theta_n)})^{1/2}(\nabla \ln\frac{\alpha_m}{\alpha_n} + 3\nabla\frac{\ln(m_n\theta_n)}{\ln(m_m\theta_m)}) \\ & + 3(\frac{2m_m^3 m_n^3\theta_m\theta_n}{\pi(m_m^2\theta_m + m_n^2\theta_n)})^{1/2}(\frac{m_n\theta_m\theta_n}{\theta_m + \theta_n})(\frac{\nabla\theta_m}{\theta_m^2} - \frac{\nabla\theta_n}{\theta_n^2}) + 6m_n(\frac{2m_m^3 m_n^3\theta_m\theta_n}{m_m^2\theta_m + m_n^2\theta_n})^{3/2}(\frac{\nabla\theta_m}{m_m\theta_m^2} - \frac{\nabla\theta_n}{m_n\theta_n^2})]\Big\} \\ & + \alpha_m k_{dil,m}\nabla\theta_m \end{aligned} \tag{T14}$$

$$k_{dil,m} = \frac{75\sqrt{\pi}}{384} d_m\rho_m\theta_{av}^{1/2} \tag{T15}$$

(f) Collision energy dissipation of component m

$$\gamma_m = \sum_{n=1}^{M}\left\{\frac{3}{d_{nm}}\left(\frac{2m_o^2\theta_m\theta_n}{\pi(m_m^2\theta_m + m_n^2\theta_n)}\right)^{1/2} - \frac{3m_o(m_m\theta_m + m_n\theta_n)}{4(m_m^2\theta_m + m_n^2\theta_n)}\nabla\cdot\mathbf{v}_m\right\}(1-e_{nm})p_m \tag{T16}$$

(g) Momentum transfer coefficient between gas and particle component m

$$\begin{aligned} \beta_{gm} = \phi & \left(\frac{150\alpha_m(1-\alpha_g)\mu_g}{\alpha_g d_m^2} + \frac{1.75\alpha_m\rho_g|v_m - v_g|}{d_m}\right) \\ & + (1-\phi)\frac{3C_D\alpha_m\alpha_g\rho_g|v_m - v_g|}{4d_m}\alpha_g^{-2.65} \end{aligned} \tag{T17}$$

(h) Particles momentum transfer coefficient between the component $n$ and $m$

$$\begin{aligned} \beta_{nm} = p_{nm}\Bigg\{ & \frac{3}{d_{nm}}\left[\frac{2(m_m^2\theta_m + m_n^2\theta_n)}{\pi m_o^2\theta_m\theta_n}\right]^{1/2} + \frac{1}{|v_m - v_s|}\Bigg[\nabla\ln\frac{\alpha_m}{\alpha_n} + 3\nabla\left(\frac{\ln(m_n\theta_n)}{\ln(m_m\theta_m)}\right) \\ & + \frac{\theta_m\theta_n}{\theta_m + \theta_n}\left(\frac{\nabla\theta_m}{\theta_m^2} - \frac{\nabla\theta_n}{\theta_n^2}\right) + \frac{5}{3}\frac{m_m m_n\theta_m\theta_n}{m_m^2\theta_m + m_n^2\theta_n}\left(\frac{m_n\nabla\theta_m}{m_m\theta_m^2} - \frac{m_m\nabla\theta_n}{m_n\theta_n^2}\right)\Bigg]\Bigg\} \end{aligned} \tag{T18}$$

## References


ANSYS Inc. (2005). FLUENT User's Guide. Canonsburg PA

Bagnold, R. A. (1954). Experiments on a gravity-free dispersion of large solid spheres in a Newtonian fluid under shear. *Proceedings of the Royal Society of London, 225*, 49–63.

Campbell, C. S., & Wang, D. G. (1991). Particle pressure in gas fluidized beds. *Journal of Fluid Mechanics, 227,* 495–508.

Chapman, S., & Cowling, T. G. (1961). *The mathematical theory of non-uniform gases.* Cambridge UK: Cambridge at the University Press.

Dan, S., Shuyan, W., Huilin, L., Zhiheng, S., Xiang, L., Shuai, W., et al. (2009). A second-order moment method of dense gas-solid flow for bubbling fluidization. *Chemical Engineering Science, 64,* 5013–5027.

Ding, J., & Gidaspow, D. (1990). A bubbling fluidization model using kinetic theory of granular flow. *AIChE Journal, 36*(4), 523–538.

Enskog, D. (1922). Kinetische theorie der Wărmeleitung, Reibung und Selbstdiffusion in gewissen verdichteten Gasen und Flŭussigkeiten. *Kungl. Sv. Vetenskapsakademiens Handl, 63,* 1–44.

Farrell, M., Lun, C. K. K., & Savage, S. B. (1986). A simple kinetic theory for granular flow of binary mixtures of smooth, inelastic spherical particles. *Acta Mechanica, 63,* 45–60.

Garzo, V., & Dufty, J. W. (2002). Hydrodynamics for a granular binary mixture at low density. *Physics of Fluids, 14*(4), 1476–1490.

Garzo, V., Dufty, J. W., & Hrenya, C. M. (2007). Enskog theory for polydisperse granular mixtures. I Navier-Stokes order transport. *Physical Review E, 76,* 03130301–03130327.

Gidaspow, D. (1994). *Multiphase flow and fluidization.* Continuum and kinetic theory description: San Diego California USA, Academic Press.

Gidaspow, D., & Huilin, L. (1997). Liquid-solid fluidization using kinetic theory. *AIChE Symposium Series: Advances in Fluidization and Fluid-Particle Systems*, 93, 12–17

Gidaspow, D., & Huilin, L. (1998). Equation of state and radial distribution functions of FCC particles in a CFB. *AIChE Journal, 44*, 279–291.

Grad, H. (1949). On the kinetic theory of rarified gases. *Communications on Pure and Applied Mathematics, 2,* 331–407.

Huang, L., Juhui, C., Guodong, L., Huilin, L., Dan, L., & Feixiang, Z. (2013). Simulated second-order moments of clusters and dispersed particles in riser. *Chemical Engineering Science, 101,* 800–812.

Huilin, L. (2017). *Kinetic theory for dense fluid-solid flow.* Science Press

Huilin, L., Gidaspow, D., & Manger, E. (2001). Kinetic theory of fluidized binary granular mixtures. *Physical Review E, 64*(6), 061301.

Huilin, L., Shuyan, W., Jianxiang, Z., Gidaspow, D., Jianmin, D., & Xiang, L. (2010). Numerical simulation of flow behavior of agglomerates in gas-cohesive particles fluidized beds using agglomerates-based approach. *Chemical Engineering Science, 65,* 1462–1473.

Huilin, L., Wenti, L., Rushan, B., Lidan, Y., & Gidaspow, D. (2000). Kinetic theory of fluidized binary granular mixtures with unequal granular temperature. *Physica a: Statistical Mechanics and Its Applications, 284,* 265–276.

Huilin, L., Yurong, H., Wentie, L., Jianmin, D., Gidaspow, D., & Bouillard, J. (2004). Computer simulations of gas-solid flow in spouted beds using kinetic-frictional stress model of granular flow. *Chemical Engineering Science, 59,* 865–878.

Iddir, H., Arastoopour, H., & Hrenya, C. M. (2005). Analysis of binary and ternary granular mixtures behavior using the kinetic theory approach. *Powder Technology, 151,* 117–125.

Jenkins, J. T., & Mancini, F. (1989). Kinetic theory for binary mixtures of smooth, nearly elastic spheres. *Physics of Fluids a: Fluid Dynamics, 1,* 2050–2057.

Jenkins, J. T., & Richman, M. W. (1985). Kinetic theory for plane flows of a dense gas of identical, rough, inelastic, circular disks. *Physics of Fluids, 28,* 485–494.

Jenkins, J. T., & Savage, S. B. (1983). Theory for the rapid flow of identical, smooth, nearly elastic, spherical particles. *Journal of Fluid Mechanics, 130,* 187–202.

Johnson, K. L. (1985). *Contact Mechanics.* Cambridge UK: Cambridge University Press.

Johnson, P. C., & Jackson, R. (1987). Frictional-collisional constitutive relations for granular materials, with application to plane shearing. *Journal of Fluid Mechanics, 176,* 67–93.

Johnson, P. C., Nott, P., & Jackson, R. (1990). Frictional-collisional equations of motion for particulate flows and their application to chutes. *Journal of Fluid Mechanics, 210,* 501–535.

Juhui, C., Shuyan, W., Dan, S., Huilin, L., Gidaspow, D., & Hongbin, Y. (2012). A second-order moment method applied to gas-solid risers. *AIChE Journal, 58,* 3653–3675.

Kim, H., & Arastoopour, H. (2002). Extension of kinetic theory to cohesive particle flow. *Powder Technology, 122,* 83–94.

Liyan, S., Weiguo, X., Guodong, L., Dan, S., Huilin, L., Yanjia, T., & Dan, L. (2012). Prediction of flow behavior of particles in a tapered bubbling fluidized bed using a second-order moment-frictional stresses model. *Chemical Engineering Science, 84,* 170–181.

Liyan, S., Weiguo, X., Huilin, L., Guodong, L., Qinghong, Z., Qing, T., & Tianyu, Z. (2015). Simulated configurational temperature of particles and a model of constitutive relations of rapid-intermediate-dense granular flow based on generalized granular temperature. *International Journal of Multiphase Flow, 77,* 1–18.

Lun, C. K. K., & Savage, S. B. (1987). A simple kinetic theory for granular flow of rough, inelastic, spherical particles. *Journal of Applied Mechanics, 54,* 47–53.

Lun, C. K. K., Savage, S. B., Jeffrey, D. J., & Chepurniy, N. (1984). Kinetic theories for granular flow: Inelastic particles in Couette flow and slightly inelastic particles in a general flow field. *Journal of Fluid Mechanics, 140,* 223–256.

Mansoori, G. A., Carnahan, N. F., Starling, K. E., & Leland, T. W., Jr. (1971). Equilibrium thermodynamic properties of the mixture of hard spheres. *the Journal of Chemical Physics, 54,* 1523–1525.

Ogawa, S., Umemura, A., & Oshima, N. (1980). On the equations of fully fluidized granular materials. *Journal of Applied Mathematics and Physic, 31,* 483–493.

Savage, S. B., & Jeffrey, D. J. (1981). The stress tensor in a granular flow at high shear rates. *Journal of Fluid Mechanics, 110,* 255–272.

Schaeffer, D. G. (1987). Instability in the evolution equations describing incompressible granular flow. *Journal of Differential Equations, 66,* 19–50.

Shuai, W., Zhenhua, H., Huilin, L., Guodong, L., Jiaxing, W., & Pengfei, X. (2012a). A bubbling fluidization model using kinetic theory of rough spheres. *AIChE Journal, 58,* 440–455.

Shuai, W., Zhenhua, H., Huilin, L., Yunchao, Y., Pengfei, X., & Guodong, L. (2012b). Hydrodynamic modeling of particle rotation in bubbling gas-fluidized beds. *International Journal of Multiphase Flow, 39,* 159–178.

Shuyan, W., Xiang, L., Huilin, L., Long, Y., Dan, S., Yurong, H., & Yonglong, D. (2009). Numerical simulations of flow behavior of gas and particles in spouted beds using frictional-kinetic stresses model. *Powder Technology, 196,* 184–193.

Songprawat, S., & Gidaspow, D. (2010). Multiphase flow with unequal granular temperatures. *Chemical Engineering Science, 65,* 1134–1143.

Srivastava, A., & Sundaresan, S. (2003). Analysis of a frictional-kinetic model for gas-particle flow. *Powder Technology, 129,* 72–85.

Strumendo, M., Gidaspow, D., & Canu, P. (2005). *Method of moments for gas-solid flows: Application to the riser* (pp. 936–942). New York, Intern Acad Pub: Circulating Fluidized Bed Technology VIII.

Syamlal, M., Rogers, W., & O'Brien, T. J. (1993). MFIX documentation theory guide, DOE/METC-94/1004 NTIS/DE94000087 U.S. Department of Energy, Office of Fossil Energy, Morgantown Energy Technology Center Morgantown, WV, Springfield, VA.

# Chapter 3
# Homogeneous and Nonhomogeneous Interfacial Momentum Closure

## Notation

| | |
|---|---|
| $a$ | Acceleration, m/s$^2$ |
| $A$ | Constant |
| $B$ | Constant |
| $C_D$ | Drag coefficient |
| $d_p$ | Particle diameter, m |
| $d_c$ | Cluster diameter, m |
| $D$ | Riser internal diameter, m |
| $e$ | Restitution coefficient |
| $E$ | Kinetic energy, m$^2$/s$^2$ |
| $E_{s,d}$ | Dispersed particles kinetic energy, m$^2$/s$^2$ |
| $E_{s,c}$ | Clusters kinetic energy, m$^2$/s$^2$ |
| $E_{s,d}$ | Dispersed particles kinetic energy, m$^2$/s$^2$ |
| $f$ | Volume fraction of dense phase |
| $f_g$ | Gas-wall friction factor |
| $\mathrm{f_p}$ | Particle-wall friction factor |
| $\boldsymbol{F}$ | Force acting on each particle or cluster, N |
| $Fr$ | Froude number |
| $g$ | Gravity, m/s$^2$ |
| $g_o$ | Radial distribution function at contact |
| $G$ | Weight function |
| $G_s$ | Solids mass flux, kg/m$^2$s |
| H | Riser height, m |
| $k_s$ | Conductivity of fluctuating energy, kg/m·s |
| $n$ | Number of particles in the dense phase per unit volume |
| $n_d$ | Number of particles in the dilute phase per unit volume |
| $N_{df}$ | Rate of energy dissipation per unit mass, W/kg |
| $p$ | Gas pressure, Pa |
| $p_s$ | Particle pressure, Pa |
| $Re$ | Reynolds number |



H. Lu et al., *Computational Fluid Dynamics and the Theory of Fluidization*,
https://doi.org/10.1007/978-981-16-1558-0_3

| | |
|---|---|
| $u_g$ | Gas velocity, m/s |
| $u_{mf}$ | Minimum fluidization velocity of particles, m/s |
| $u_s$ | Particle velocity, m/s |
| U | Superficial gas velocity, m/s |
| $U_c$ | Superficial slip velocity in dense phase, m/s |
| $U_d$ | Superficial slip velocity in dilute phase, m/s |
| $U_i$ | Superficial slip velocity of interface, m/s |
| $V$ | Volume of dense phase in the control volume, $m^3$ |
| $V_d$ | Volume of dilute phase in the control volume, $m^3$ |
| $v_g$ | Gas velocity, m/s |
| $v_s$ | Solids velocity, m/s |
| $x$ | Transverse distance from axis, m |
| z | Height along vertical direction, m |

**Greek Letters**

| | |
|---|---|
| $\beta$ | Drag coefficient, $kg/m^3$ s |
| $\gamma$ | Collisional energy dissipation, $kg/m{\cdot}s^3$ |
| $\varepsilon_d$ | Porosity in the dilute phase |
| $\varepsilon_c$ | Porosity in the dense phase |
| $\alpha_g$ | Porosity |
| $\alpha_s$ | Solids volume fraction |
| $\alpha_{max}$ | Maximum porosity for particle aggregating |
| $\theta$ | Granular temperature, $m^2/s^2$ |
| $\mu_g$ | Gas shear viscosity, kg/m s |
| $\mu_s$ | Solids shear viscosity, kg/m·s |
| $\rho_g$ | gas density, $kg/m^3$ |
| $\rho_s$ | Density of solids phase, $kg/m^3$ |
| $\tau_g$ | Gas stress tensor, Pa |
| $\tau_s$ | Particle stress tensor, Pa |
| $\Omega$ | Energy dissipation rate, $W/m^3$ |
| $\Omega_{co}$ | Collisional energy dissipation rate, W/kg |
| $\Omega_{dr}$ | Rate of energy dissipation of drag force, W/kg |
| $\Omega_{vi}$ | Viscous energy dissipation, W/kg |

**Subscripts**

| | |
|---|---|
| c | Dense phase with clusters |
| d | Dilute phase |
| g | Gas phase |
| s | Solids phase |
| mf | Minimum fluidization. |

## 3.1 Introduction

For simulation, design, and scale-up of commercial fluidized beds and gas–particle flow systems, a computationally feasible approach may be obtained from the averaged continuum equations of motion for both fluid and particles, which is often called two-fluid model (TFM). The continuum approach generally relies on closures for the drag forces between the fluid phase and the solids phase. The averaging process leading to the TFM equations erases the details of flow at the level of individual particles; but their consequences appear in the averaged equations through terms such as drag force, for which one must develop constitutive relations. For example, in the momentum balance equations, constitutive relations are needed for the gas–particle interaction forces. Gas–particle flows in fluidized beds and riser reactors are inherently oscillatory, and they manifest in nonhomogeneous structures over a wide range of length and time scales. Thus, if one sets out to solve the microscopic TFM for gas–particle fluidized beds, the grid sizes of the order of 10-particle diameter or smaller become essential. Moreover, such fine spatial resolution reduces the time steps required, further increasing the needed computational time. For most devices of practical (commercial) interest, it is nearly impossible to resolve all heterogeneous flow structures in large-scale industrial risers using a computational grid size of the order of a few particle diameters. In addition, such extremely fine spatial grids and small time steps are unaffordable and require significant computational time and use of significant computational facilities.

Gas-particle flows in large fluidized beds and risers often are simulated by solving discretized versions of the TFM equations over a coarse spatial grid. Such coarse grid simulations do not resolve the small-scale (i.e., subgrid scale) spatial structures in fluidized beds. The effect of the large-scale structures using coarse grids must be accounted for through appropriate modifications to the closures. One of the major sources of numerical inaccuracy for the TFM originates from the models used for the calculation of fluid-particles drag force. The homogeneous drag models, including Gidaspow (1994) A and B drag models, Wen and Yu (1966) drag coefficient, Syamlal and O'Brien (1987) drag coefficient model and Huilin-Gidaspow (Huilin et al. 2003) drag models, assume a homogeneous structure inside the control volumes, which is not valid due to the formation of clusters (dense phase) in the concentrated particulate phase (e.g., $\alpha_s > 2\%$). The effective drag coefficient in the coarse grid simulations will be lower than that in the homogeneous TFM to reflect the tendency of the gas to flow more easily around the clusters (bypass the clusters) than through a homogeneous distribution of the particles. Qualitatively, this is equivalent to an effectively larger apparent size for the particles. Therefore, any coarse grid continuum simulation of gas–solid fluidized beds should include subgrid corrections to the homogeneous drag force acting on the particles.

There are at least two approaches, the filtered two-fluid models (fTFM) and the heterogeneous structure-based drag models, used for modeling flow of heterogeneous structures of clusters and bubbles in fluidized beds. The conservation equations of the fTFM are derived by applying space filtering operators, assuming the flow of heterogeneous structures dissolved into the spatially resolved parts and unresolved parts (Agrawal et al. 2001). The resolved parts are modeled using kinetic theory-based TFM, while the unresolved parts are solved using filtered sub-grid constitutive models. The constitutive relations of the filtered drag force and solid stress proposed by Igci et al. (2008) are solved as a function of filtered solids volume fraction, Froude number and dimensionless filter length based on the terminal settling velocity of particles (Igci and Sundaresan 2011; Sarkar et al. 2016).

The heterogeneous structure-based drag models, including energy minimization multiscale (EMMS) method (Li and Kwauk 2003) and cluster structure-dependent (CSD) drag model (Shuai et al. 2011), assume that the gas-particle mixture consists of the heterogeneous structures with the particle-rich dense phase as the form of clusters and the gas-rich dilute phase as the form of dispersed particles. The EMMS method assumes that the suspending and transporting energyis minimum, and is known as a stability condition (Li and Kwauk 2003). The drag forces are solved by the set of equations with the stability condition. The drag coefficient is expressed by heterogeneity index which is correlated with solids volume fraction only based on Wen and Yu empirical correlation (Wen and Yu 1966). The CSD drag model predicts drag coefficient of heterogeneous structures using a set of nonlinear conservation equations (Shuai et al. 2014; Dan et al. 2018). The cluster diameter is solved from the equation set without additional empirical or analytical correlations.

## 3.2 Governing Equations of Gas-Particles KTGF-Based TFM Model

The two-fluid model (TFM) is formulated applying the Eulerian continuum concept for the continuous phase, while the governing equations of the solids phase are developed in accordance with the principles of kinetic theory of granular flow (KTGF). In this theory it is postulated that the particulate system can be represented considering a collection of identical, smooth, rigid spheres.

In the KTGF-based TFM model (Gidaspow 1994; Huilin 2017), the mass conservation equations of gas phase ($i = g$) and solids phases ($i = s$) are expressed as

$$\frac{\partial}{\partial t}(\alpha_i \rho_i) + \nabla \cdot (\alpha_i \rho_i \mathbf{v}_i) = 0 \tag{3.1}$$

The momentum conservation equations of gas phase and solids phase are expressed as

$$\frac{\partial}{\partial t}(\alpha_g \rho_g \mathbf{v}_g) + \nabla \cdot (\alpha_g \rho_g \mathbf{v}_g \mathbf{v}_g) = -\alpha_g \nabla p + \alpha_g \nabla \cdot \tau_g + \beta_{gs}(v_s - v_g) + \alpha_g \rho_g \mathbf{g} \tag{3.2}$$

$$\frac{\partial}{\partial t}(\alpha_s \rho_s \mathbf{v}_s) + \nabla \cdot (\alpha_s \rho_s \mathbf{v}_s \mathbf{v}_s) = -\alpha_s \nabla p - \nabla p_s + \nabla \cdot \tau_s + \beta_{gs}(v_g - v_s) + \alpha_s \rho_s \mathbf{g} \tag{3.3}$$

where $\alpha_i$ is volume fractions of gas and solids phase, $v_i$ is velocity vectors of gas and solids phases, $p$ is pressure and $\beta_{gs}$ is the interface momentum exchange coefficient.

The conservation of particles fluctuating energy for granular temperature is (Syamlal et al. 1993; Gidaspow 1994; Fox 2014):

$$\frac{3}{2}\left[\frac{\partial}{\partial t}(\alpha_s \rho_s \theta) + \nabla \cdot (\alpha_s \rho_s \theta)\mathbf{v}_s\right] = (-\nabla p_s \mathbf{I} + \tau_s) : \nabla \mathbf{v}_s + \nabla \cdot (k_s \nabla \theta) - \gamma_s + \phi_s \tag{3.4}$$

where $\gamma_s$ is the collisional energy dissipation, and $\phi_s$ is the granular energy exchange between phases.

From Eq. (3.1)–(3.4), the control volume (CV) parameters, including gas volume fraction $\alpha_g$, gas pressure $p$ (or $\Delta p/\Delta z$), granular temperature $\theta$, and velocities of gas phase $v_g$ and solids phase $v_s$, are solved from KTGF-based TFM models. Hence, the homogenous flow or heterogeneous flow of gas and particles suspension relates with CV parameters.

## 3.3 Homogeneousfluid-Particle Drag

In order to solve the momentum conservation equations of gas and solids phases, one would need closure for the drag force or momentum transfer between the gas phase and the solids phase. The drag coefficients are usually obtained from experimental correlations, primarily based on two types of experimental data. One type, valid for high values of the solids volume fraction, is the packed-bed pressure drop data. The other type of experimental data is the terminal velocity of fluidized beds or settling beds.

### *3.3.1 Gidaspow Drag Coefficient Model*

The fluid-particle interaction closures applied in the momentum balances originate from the work on the Newton's secondlaw for particles. A one-dimensional steady state gas momentum balance for a constant cross-sectional area is

$$\frac{d(\rho_g \alpha_g v_g^2)}{dz} = -\alpha_g \frac{dp}{dz} - \rho_g g \alpha_g - \beta_A (v_g - v_s) - \frac{4\tau_{wg}}{D_t} \tag{3.5}$$

The third term on the right-hand side represents the rate of momentum dissipation due to relative motion. The last term is the rate of momentum dissipation due to wall shear with the diameter $D_t$ and wall shear $\tau_{wg}$. With no acceleration, wall friction, or gravity, the gas momentum balance, Eq. (3.5), is

$$-\alpha_g \frac{dp}{dz} - \beta_A (v_g - v_s) = 0 \tag{3.6}$$

The friction coefficient is obtained from the Ergun equation (Ergun 1952):

$$\frac{\Delta p}{\Delta z} = 150 \frac{\alpha_s^2}{\alpha_g^3} \frac{\mu_g U_0}{d_p^2} + 1.75 \frac{\alpha_s \rho_g U_0^2}{\alpha_g^3 d_p} \tag{3.7}$$

where $U_0$ is the superficial velocity $U_0 = \alpha_g(v_g - v_s)$, A comparison of Eqs. (3.6) and (3.7) shows that the drag for model A is

$$\beta_A = 150 \frac{\alpha_s^2 \mu_g}{\alpha_g d_p^2} + 1.75 \frac{\alpha_s \rho_g |v_g - v_s|}{d_p} \qquad \alpha_g \leq 0.8 \tag{3.8}$$

We see that the drag coefficient $\beta_A$ divided by the viscosity is the reciprocal permeability. The value of the drag coefficient is proportional to the square of the particle diameter for low flows.

The gas momentum balance is also obtained by subtracting the dispersed phase momentum balance from the mixture momentum balance.

$$\rho_g \alpha_g v_g \frac{dv_g}{dz} = -\frac{dp}{dz} - \rho_g g \alpha_g - \beta_B (v_g - v_s) - \frac{4\tau_{wg}}{D_t} \tag{3.9}$$

The momentum rate due to fluid pressure acting on the fluid is different from the term that the porosity multiplies the gradientof pressure in the model in which the pressure drop is taken to be in the gas and solids phases. The relation of drag coefficient can be obtained for the case of zero velocity gradient, zero gravity, and zero wall friction. The drag coefficient $\beta_B$ of model B is

$$\beta_B = \frac{\beta_A}{\alpha_g} \tag{3.10}$$

For dilute suspensions, Wen and Yu (1966) have extended the works of Richardson and Zaki (1954) to derive an expression for pressure drop prediction in a particulate bed. For porosities greater than 0.8, if $\Delta p/\Delta x$ is replaced by Wen and Yu's expression for pressure drop, the drag coefficient in this porosity range becomes

$$\beta_A = \frac{3}{4} C_D \frac{\alpha_s \alpha_g \rho_g |v_g - v_s|}{d_p} f(\alpha_g) \qquad \alpha_g > 0.8 \tag{3.11}$$

where $C_D$, the drag coefficient, is related to the Reynolds number by Rowe (1961)

$$C_D = \frac{24}{(\mathrm{Re}_s)}[1 + 0.15(\mathrm{Re}_s)^{0.687}] \qquad (\mathrm{Re}_s) < 1000 \tag{3.12}$$

$$C_D = 0.44 \qquad (\mathrm{Re}_s) \geq 1000 \tag{3.13}$$

$$(\mathrm{Re}_s) = \frac{\alpha_g \rho_g |v_g - v_s| d_p}{\mu_g} \tag{3.14}$$

In Eq. (3.11) $f(\alpha_g)$ shows the effect due to the presence of other particles in the fluid and acts as a correction to the usual Stokes law for free fall of a single particle. Gidaspow and Ettehadieh (1983) have used

$$f(\alpha_g) = \alpha_g^{-2.65} \tag{3.15}$$

### 3.3.2 *Huilin-Gidaspow Drag Coefficient Model*

The Huilin-Gidaspow drag model (Huilin and Gidaspow 2003), known as Gidaspow-blend drag model (Estejab and Battaglia 2016), has extended to avoid switching between the Ergun equation and the Wen-Yu drag relation. The difference between the two models lies in an equation that is used to combine the results of the Ergun and Wen-Yu drag models to "blend" or smooth the transition when the particle characteristics transition from one regime (dilute) to the other (dense). The Huilin-Gidaspow drag model expresses as follows (Huilin et al. 2003)

$$\beta_A = (1 - \phi)\beta_{\mathrm{A(Ergun)}} + \phi\beta_{A(\mathrm{Wen-Yu})} \tag{3.16}$$

where the blending function, $\phi$, is

$$\phi = \{\arctan[150 \times 1.75(0.2 - \alpha_s)]/\pi\} + 0.5 \tag{3.17}$$

### 3.3.3 *Syamlal and O'Brien Drag Coefficient Model*

A formula for the fluid-solids drag coefficient for a multiparticle system is derived from a Richardson-Zaki type velocity-voidage correlation (1954) according to the Garside and Al-Dibouni (1977) equations (Syamlal and O'Brien 1987). The drag coefficient is expressed as a function of multiparticle drag coefficient $C_D$

$$\beta_A = \frac{3}{4}\frac{\alpha_s\alpha_g\rho_g}{V_r^2 d_p}C_D|v_g - v_s| \tag{3.18}$$

where $V_r$ is the ratio between the falling velocity of asuspension (superficial) and the terminal velocity of a single particle.

$$V_r = 0.5\left(A - 0.06\text{Re} + \sqrt{(0.06\text{Re})^2 + 0.12\text{Re}(2B - A) + A^2}\right) \tag{3.19}$$

$$A = \alpha_g^{4.14} \tag{3.20}$$

$$B = \begin{cases} 0.8\alpha_g^{1.28} & \alpha_g \le 0.85 \\ \alpha_g^{2.65} & \alpha_g > 0.85 \end{cases} \tag{3.21}$$

and the Reynolds number is given by

$$\text{Re} = \frac{\rho_g|v_g - v_s|d_p}{\mu_g} \tag{3.22}$$

The single-sphere drag function is

$$C_D = \left(\frac{0.63}{\text{V}_\text{r}} + \frac{4.8}{\sqrt{\text{V}_\text{r}\text{Re}}}\right)^2 \tag{3.23}$$

According to Garside and Al-Dibouni formula, the multiparticle drag coefficient $C_D$ and its derivative with respect to voidage $\alpha_g$ are continuous, except at the voidage of 0.85 where $C_D$ is continuous, but its derivative is discontinuous.

### 3.3.4 Gibilaro Drag Coefficient Model

The pressure drops were correlated with solids volume fractions on the basis of theoretical considerations according to experimental measurements of the fixed bed. It was used to produce drag coefficient correlations for individual particles in the bed which in turn yield a general expression for the drag force on a particle in a fluidized suspension (Gibilaro et al. 1985).

$$\beta_A = \left(\frac{18}{\text{Re}} + 0.33\right)\frac{\alpha_s\rho_g}{d_p}|v_g - v_s|\alpha_g^{-1.8} \tag{3.24}$$

with the Reynolds number as

$$\mathrm{Re} = \frac{\alpha_g \rho_g |v_g - v_s| d_p}{\mu_g} \tag{3.25}$$

## 3.4 Filtered or Subgrid Model

The filtered two-fluid model (fTFM) equations are coarse-grained through a filtering operation that amounts to spatial averaging over some chosen filter length scale. In these filtered (coarse grained) equations, the consequences of the flow structures occurring on a scale smaller than a chosen filter size appear through residual correlations for which one must derive or postulate constitutive models (Igci et al. 2008). If the several assumptions innate to the filtering methodology hold true, the filtered equations should produce a solution with the same macroscopic features as the finely resolved kinetic theory model solution.

According to the method proposed by Igci et al. (2008), if $\alpha_s(x, t)$ denotes the particle volume fraction at location $x$ and time $t$ is obtained by solving the microscopic TFM, the filtered particle volume fraction can be defined as

$$\overline{\alpha}_s(\boldsymbol{x}, t) = \int\limits_V G(\boldsymbol{x} - z)\alpha_s(z, t)dz \tag{3.26}$$

where $G(\boldsymbol{x}–z)$ is a weight function that depends on $\boldsymbol{x}$ and $z$, and $V$ denotes the region over which the gas–particle flow occurs. The weight function satisfies

$$\int G(\boldsymbol{x} - z)dz = 1.0 \tag{3.27}$$

By choosing how rapidly $G(\boldsymbol{x}–z)$ decays with distance measured from $\boldsymbol{x}$, one can change the filter size. The box filter kernel is defined as

$$G(\boldsymbol{x}, z) = \begin{cases} \frac{1}{\Delta_{fil}^3} & |\boldsymbol{x} - z| \leq \Delta_{fil}/2 \\ 0 & \text{otherwise} \end{cases} \tag{3.28}$$

where $\Delta_{fil}$ is the filter size. Igci et al. (2008) defined the fluctuation in particle volume fraction as

$$\alpha_s' = \alpha_s - \overline{\alpha}_s \tag{3.29}$$

The filtered phase velocities are defined according to

$$\overline{\alpha}_s(\boldsymbol{x},t)\overline{v}_s(\boldsymbol{x},t) = \int_V G(\boldsymbol{x}-\boldsymbol{z})\alpha_s(\boldsymbol{z},t)v_s(\boldsymbol{z},t)d\boldsymbol{z} \tag{3.30}$$

$$\overline{\alpha}_g(\boldsymbol{x},t)\overline{v}_g(\boldsymbol{x},t) = \int_V G(\boldsymbol{x}-\boldsymbol{z})\alpha_g(\boldsymbol{z},t)v_g(\boldsymbol{z},t)d\boldsymbol{z} \tag{3.31}$$

where $v_g$ and $v_s$ denote local gas- and particle-phase velocities appearing in the microscopic TFM. Thus, the filtered two fluid model (fTFM) equations have been deduced by filtering the TFM equations using a filter. The fTFM equations are more suitable for numerical integration at a coarser mesh resolution; however, because of lack of scale separation, the constitutive models to be employed in conjunction with fTFM can account for the unresolved subfilter-scale structures through scale-dependent subfilter models. (This is conceptually similar to large eddy simulation of single-phase turbulent flows). Solving the fTFM numerically, it is commonly assumed that the filter size is the same as the grid size. The algebraic form for the filtered drag coefficient is

$$\beta_{fil} = \frac{32Fr_f^{-2} + 63.02Fr_f^{-1} + 129}{Fr_f^{-3} + 133.6Fr_f^{-2} + 66.61Fr_f^{-1} + 129}\beta_{micro} \tag{3.32}$$

where $\beta_{micro}$ is the drag coefficient in the microscopic TFM. $Fr_f$ is the Froude number based on the filter size and is defined as

$$Fr_f = \frac{u_t^2}{g\Delta_f} \tag{3.33}$$

where $u_t$ is the terminal settling velocity of particles and $\Delta_f$ is the filter size. For further discussions on dependence of the residual correlations on the filtersize, filtered particle volume fraction, and filtered slip velocity, all of which serve as markers for the extent of subfilter-scale in-homogeneity (Ozarkar et al. 2015; Sarkar et al. 2016).

## 3.5 Energy Minimization Multi-scale (EMMS) Model

The energy minimization multi-scale (EMMS) model was proposed by Li and Kwauk (2003) based on the coexistence of both dense and dilute regions in a riser. The multi-scale gas–solid interactions can be quantified by the drag coefficients corresponding to the three subsystems of the dilute phase and dense phase and the interface. The three scale interactions include the micro-scale interaction between gas and solid which is correlated with the size of the constituent particles and is present in both the dense and dilute phases, meso-scale interaction between the dense clusters and dilute phase with the size of clusters, and macro-scale interaction between the global gas–solid system and its boundaries. Eight parameters are needed to describe the state of a system.These parameters are found by minimization of the mass-specific energy consumption for suspending and transporting the particles as the stability criteria for flow structure inside the reactor. The EMMS method assumes that the suspending and transporting energy is minimum, and is known as a stability condition.

$$N_{st} = \frac{\rho_s - \rho_g}{\rho_s} g \left\{ U_g + (fU_c - U_g) \frac{(\varepsilon_d - \varepsilon_c) f^2}{1 - \varepsilon_d + (\varepsilon_d - \varepsilon_c) f} \right\} \rightarrow \min \quad (3.34)$$

where $\varepsilon_c$ and $\varepsilon_d$ are the gas volume fractions of the dense phase and dilute phase. The drag forces are solved by the set of equations with the stability condition. The drag coefficient is expressed by heterogeneity index which is correlated with solids volume fraction based on Wen and Yu empirical correlation (Wen and Yu 1966). The drag expression is expressed as

$$\beta = \begin{cases} \frac{3}{4} \frac{(1-\alpha_g)\alpha_g}{d_p} \rho_g |v_g - v_s| C_d \omega & \alpha_g > \alpha_g^* \\ 150 \frac{\alpha_s^2 \mu_g}{\varepsilon_g d_p^2} + 1.75 \frac{\alpha_s \rho_g |v_g - v_s|}{d_p} & \alpha_g \le \alpha_g^* \end{cases} \quad (3.35)$$

where $\omega$ is the heterogeneity factor. For the specific operating conditions (air-fluid catalytic cracking particle system, $\rho_s = 930$ kg/m$^3$, $G_s = 14.3$ kg/(m$^2$ s) and $U_g = 1.52$ m/s), the heterogeneity factor could be written as

$$\omega = \begin{cases} -0.5760 + \frac{0.0214}{4(\alpha_g - 0.7463)^2 + 0.0044} & \alpha_g^* < \alpha_g \le 0.82 \\ -0.0101 + \frac{0.0038}{4(\alpha_g - 0.7789)^2 + 0.0040} & 0.82 < \alpha_g \le 0.97 \\ -31.8295 + 32.8295\alpha_g & \alpha_g > 0.97 \end{cases} \quad (3.36)$$

where $\alpha_g$* is the gas volume fraction at which $\omega$ equals the correction factor of Gidaspow (1994).

## 3.6 Dynamic Cluster Structure-Dependent Drag Model

### *3.6.1 Heterogeneous Flow with Dilute Phase and Dense Phase*

The local transient solids volume fractions were measured using an optical fiber probe in a riser. The riser height and diameter are 4.5 m and 120 mm, respectively. The diameter and density of particles are 70 micron and 1650 kg/m$^3$. A sample time-trace of the transient solids volume fractions is shown in Fig. 3.1 at the center of the riser

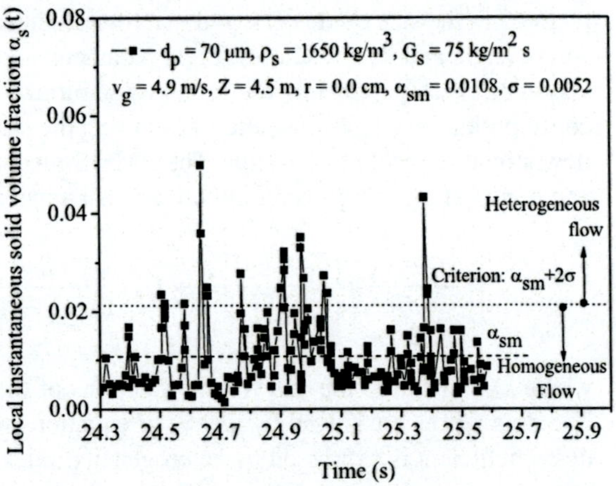


**Fig. 3.1** Measured instantaneous solids volume fractions at the center of the riser

at the superficial gas velocity and solid mass flux of 4.9 m/s and 75 kg/m$^2$s. From measured instantaneous solids volume fractions, the time-averaged solids volume fraction $\alpha_{sm}$ and its standard deviation σ are determined.

To obtain quantitative characterization of flow of clusters, Soong et al. (1995) developed criteria using statistical methodology based on measured instantaneous solids volume fractions. Sharma et al. (2000) used a similar approach as Soong et al. (1995) to study the effect of particle size and superficial gas velocity on particle cluster dynamics. Their criteria were that the solids volume fraction in a cluster must be significantly greater than the time averaged solid fraction at that local position.

$$\alpha_s(t) \geq \alpha_{sm} + n\sigma = \alpha_{cr} \tag{3.37}$$

where $\alpha_{sm}$ is the time-averaged solids volume fraction. $\sigma$ is the standard deviation. $n$ is a constant. This guideline gives a criterion that the local instantaneous solids volume fraction for a cluster must be greater than the time mean solids volume fraction by constant $n$ times the standard deviation. A cluster would thus be identified if the instantaneous solid concentration exceeds this threshold, existing until the solid fraction again drops below this threshold. The value of the constant $n$ is selected with some degree of arbitrariness. Soong et al. (1995) chooses $n = 3$, Gu and Chen (1998) set $n = 2.5$, Tuzla et al. (1998) and Sharma et al. (2000) are taken as $n = 2$. Therefore, the value of $n$ affects the values of cluster dynamics from experimental measurements. Unfortunately, the criterion of Eq. (3.37) did not use in the determination of clusters existence because of unknown standard deviation in numerical simulations. Hence, the criterion for the existence of clusters has proved to be important for characterization of particle cluster dynamics from numerical simulations.

From measured instantaneous solids volume fractions, the local gas and particles suspension flow at least resolves into two flow states, seeing Fig. 3.1. One is known

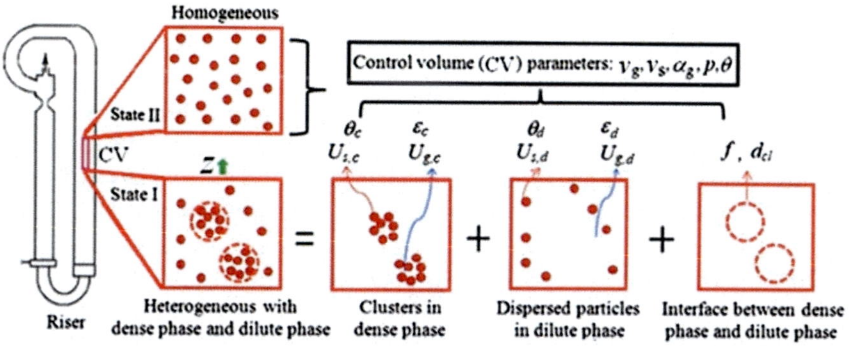


**Fig. 3.2** Homogenous and heterogeneous structures with the dense phase ($\varepsilon_c$ less than $\alpha_g$) and the dilute phase ($\varepsilon_d$ larger than $\alpha_g$) of five CV parameters and ten structure parameters in the CV

as the heterogeneous flow with clusters if the solids volume fraction $\alpha_s$ is larger than the threshold $\alpha_{cr}$. The other is the homogenous flow without clusters when $\alpha_s$ is smaller than $\alpha_{cr}$. This concept applies to a control volume (CV) for gas-particles mixture in numerical simulations. Figure 3.2 shows flow states of gas and particles mixture in a CV. The state of gas-particle flow maybe is heterogeneous structure with clusters or homogeneous without clusters in CVs. The heterogeneous structure of the state I is dissolved into a dilute phase where the gas volume fraction is larger than $\alpha_g$, a dense phase in which the gas volume fraction is less than $\alpha_g$ and the interface between the dilute phase and the dense phase in the control volume $V$. For the state I, the dense phase volume part $V_c$ consists of clusters with a mean cluster diameter $d_c$, gas superficial velocity $U_{g,c}$, clusters superficial velocity $U_{s,c}$, gas volume $V_{g,c}$ and granular temperature of particles within the clusters $\theta_c$. The dilute phase volume part $V_d$ of dispersed particles is described by the gas superficial velocity $U_{g,d}$, gas volume $V_{g,d}$, dispersed particles superficial velocity $U_{s,d}$ and dispersed particles granular temperature $\theta_d$. The dense phase volume fraction of the CV defines $f = V_c/V$. The gas volume fractions of the dense phase and the dilute phase are $\varepsilon_c = V_{g,c}/V_c$ and $\varepsilon_d = V_{g,d}/V_d$ in the CV. Thus, the heterogeneous flow of gas and particles mixture is characterized by ten structure parameters which are related to five CV parameters.

### 3.6.2 *Criterion for Identification of Clusters*

When the heterogeneous structure exists in the CV, it is resolved into the dilute phase and the dense phase with an interphase between them. The different scale interactions include a micro-scale interaction and meso-scale interaction. The micro-scale interactions between gas and particles are correlated with the size of the constituent particles and presents in both the dense phase $F_c$ and dilute phase $F_d$. These interaction forces are expressed as the magnitude of the forces acting on a single particle in the dense and dilute phases. The meso-scale interaction $F_i$ between the dense clusters and dilute broth surrounding them is correlated with the size of clusters, indicating the identification and characterization of particle clusters must be required.

The kinetic energy equation of solids phase was derived according to the kinetic-theory-based model (Gidaspow 1994; Fox 2014). For flow of heterogeneous structures, the granular energy $E_s$ consists of the energy component of dispersed particles in the dilute phase $E_{s,d}$ and the energy component of clusters in the dense phase $E_{s,c}$ according to the velocities of the dilute phase $u_{s,d}$ and dense phase $u_{s,c}$.

$$E_{s,d} = \frac{1}{2}\mathbf{u}_{s,d} \cdot \mathbf{u}_{s,d} + \frac{3}{2}\theta_d \tag{3.38}$$

$$E_{s,c} = \frac{1}{2}\mathbf{u}_{s,c} \cdot \mathbf{u}_{s,c} + \frac{3}{2}\theta_c \tag{3.39}$$

The conservation energy equation of solids phase is expressed by Eq. (T3.5). The fifth term on the left hand side represents the granular temperature flux components of the dilute and dense phases. A similar operation is performed to analyze the gas energy $E_g$ in the dilute phase and dense phase. The conservation energy equation of gas phase is expressed by Eq. (T3.6). The fourth term on the left hand side represents gas heat fluxes of the dilute and dense phase. Then, the addition of these equations yields the energy balance of gas-particles suspension as follows:

$$\begin{aligned}
&\tfrac{\partial}{\partial t}(E_s + E_g) + \tfrac{\partial}{\partial z}[f^2(\rho_s U_{s,c} E_{s,c} + \rho_g U_{g,c} E_{g,c}) + (1-f)^2(\rho_s U_{s,d} E_{s,d} + \rho_g U_{g,d} E_{g,d})] \\
&\quad +(\alpha_g v_g + \alpha_s v_s)\tfrac{\partial p}{\partial z} + \tfrac{\partial}{\partial z}\left[\left(\tfrac{f U_{s,c}}{1-\varepsilon_c}\right)p_{s,c} + \left(\tfrac{(1-f)U_{s,d}}{1-\varepsilon_d}\right)p_{s,d}\right] + \tfrac{\partial}{\partial z}(q_{s,d} + q_{s,c}) \\
&\quad + \tfrac{\partial}{\partial z}(q_{g,d} + q_{g,c}) + [f^2 \rho_s U_{s,c} + (1-f)^2 \rho_s U_{s,d}]g + [f \rho_g U_{g,c} + (1-f)\rho_g U_{g,d}]g \\
&\quad = [n_d F_d U_d + n_c F_c U_c + n_i (1-f) F_i U_i] + (\Gamma_{s,c} + \Gamma_{s,c}) + (\Gamma_{g,d} + \Gamma_{g,c})
\end{aligned} \tag{3.40}$$

$$E_s = f(1-\varepsilon_c)\rho_s E_{s,c} + (1-f)(1-\varepsilon_d)\rho_s E_{s,d} \tag{3.41}$$

$$E_g = f \varepsilon_c \rho_g E_{g,c} + (1-f)\varepsilon_d \rho_g E_{g,d} \tag{3.42}$$

In Eqs. (3.70)–(3.42), the gas energy components of the dilute phase $E_{g,d}$ and the dense phase $E_{g,c}$ are unclosed, and must be found from the transport equations, leading to Eq. (3.40) is more complicated than in the gas energy conservation equation. But, we observe that the terms at the right hand side represent the rate of energy dissipations of gas-particles suspension through three contributions by hydrodynamic forces, interactions of collisions of particles and gas viscous forces of heterogeneous structures.

$$\Omega = [n_d F_d U_d + n_c F_c U_c + n_i F_i U_i (1-f)] + (\Gamma_{s,d} + \Gamma_{s,c}) + (\Gamma_{g,d} + \Gamma_{g,c}) \tag{3.43}$$

To keep the energy dissipation rate as low as possible, particles tend to aggregate into clusters to reduce the gas flow resistance and the collisional energy dissipations, while gas tends to bypass the clusters instead of penetrating through clusters (Gidaspow 1994; Huilin 2017). Hence, the criterion of the existence of heterogeneous structures is the energy dissipation rate per unit mass of mixture is minimum:

$$N_{df} = \frac{\Omega}{\alpha_g \rho_g + (1 - \alpha_g)\rho_s} = \Omega_{dr} + \Omega_{co} + \Omega_{vi} = \min \tag{3.44}$$

This is a criterion that the CV energy dissipation rate must be minimization when the existence of particle clusters occurs. The heterogeneous flow would thus be identified. Equation (3.44) is a extreme value equation of the rate of energy dissipation including the energy dissipation component of drag forces $\Omega_{dr}$, the collisional energy dissipation component $\Omega_{co}$ and gas viscous energy dissipation component $\Omega_{vi}$. As the flow state of gas-particles suspension of the CV satisfies the criterion of minimization energy dissipation rates by Eq. (3.44), the energy dissipation components of drag forces collisions are expressed

$$\Omega_{dr,\min} = \frac{[n_d F_d U_d + n_c F_c U_c + n_i F_i U_i (1 - f)]_{\min}}{\alpha_g \rho_g + (1 - \alpha_g)\rho_s} \tag{3.45}$$

$$\Omega_{co,\min} = \frac{(\Gamma_{s,d} + \Gamma_{s,c})_{\min}}{\alpha_g \rho_g + (1 - \alpha_g)\rho_s} \tag{3.46}$$

### 3.6.3 Dynamic Cluster Structure-Dependent (DCSD) Drag Model

For heterogeneous flow of gas and particles with clusters, the drag forces acting on a suspension of particles can be formulated in terms of the forces acting on particles in the dense and dilute phases and interaction force between the dense phase and the dilute phase from momentum balance equationsof the dense phase and dilute phase (Huilin 2017), the DCSD drag coefficient is expressed as follows:

$$\beta_{DCSD} = \frac{n_c F_{c,\min} + n_d F_{d,\min} + n_i F_{i,\min}}{|v_g - v_s|} \qquad N_{df}^{BEV} = \min \tag{3.47}$$

For homogenous flow of gas and particles mixture, the drag coefficient is calculated using Huilin-Gidaspow drag correlation with the switch function $\phi$:

$$\beta_{HG} = \phi\left(\frac{150\alpha_s(1-\alpha_g)\mu_g}{\alpha_g d_s^2} + \frac{1.75\alpha_s\rho_g|v_s - v_g|}{d_s}\right)$$
$$+ (1-\phi)\frac{3C_D\alpha_s\alpha_g\rho_g|v_s - v_g|}{4d_s}\alpha_g^{-2.65} \quad N_{df}^{BEV} \neq \min \tag{3.48}$$

From Eqs. (3.47) and (3.48), the drag coefficient in the CV is modeled as follows:

$$\beta_{gs} = \begin{cases} \beta_{DCSD} & N_{df}^{BEV} = \min \\ \beta_{HG} & N_{df}^{BEV} \neq \min \end{cases} \tag{3.49}$$

### 3.6.4 *Transient Equations of Dilute Phase and Dense Phase*

In order to solve the criterion Eq. (3.44), the forces acting on a single particle of the dense phase $F_c$ and the dilute phase $F_d$, and the force acting on a cluster through the interface $F_i$ are first evaluated in the CVs. These forces are determined from momentum balance of gas and solids phases.

For fluid catalytic cracking (FCC) particles, the solids velocity is close to the gas velocity in the direction of the flow (Grace et al. 1997; Gidaspow 1994; Huilin 2017). Thus, one-dimensional flow of heterogeneous structuresis assumed along the mean flow direction (vertical $z$ direction) in a first approximation. Although one-dimensional equations of heterogeneous structures are easier to obtain from the momentum equations of gas and solids phases, the equations are solved to be useful for further investigations of multidimensional equations of hydrodynamics of heterogeneous structures.

The one-dimensional transient momentum equation of dispersed particles is expressed by Eq. (T3.1), as shown in Table 3.1 (Huilin 2017; Dan et al. 2018). The second term on the right hand side is the drag force in the dilute phase. The forth term is the dispersed particles pressure, and the last term is momentum exchange between the dilute phase and the dense phase. Because the density of particles is greater than that of gas phase, Eq. (T3.1) can be rewritten to

$$n_d F_d = (1-f)(1-\varepsilon_d)(\rho_s - \rho_g)(g + a_{sd,l} + a_{sd,c}) + (1-f)(1-\varepsilon_d)\frac{\partial p}{\partial z}$$
$$+ \frac{\partial p_{s,d}}{\partial z} + \beta_{dc}\left(\frac{U_{s,c}}{1-\varepsilon_c} - \frac{U_{s,d}}{1-\varepsilon_d}\right) \tag{3.50}$$

where $a_{sd,l}$ and $a_{sd,c}$ are the temporal acceleration and convective acceleration of dispersed particles, and they can be deduced from the given equations. The temporal acceleration and convective acceleration represent the rate of velocity change with

**Table 3.1** Model equations of DCSD drag coefficient

(1) One-dimensional transient momentum equation of dispersed particles

$$\frac{\partial}{\partial t}[(1-f)^2\rho_s U_{s,d}] + \frac{\partial}{\partial z}\left[(1-f)^2\rho_s U_{s,d}\frac{(1-f)U_{s,d}}{(1-\varepsilon_d)}\right] = -(1-f)(1-\varepsilon_d)\frac{\partial p}{\partial z} + n_d F_d - (1-f)(1-\varepsilon_d)\rho_s g - \frac{\partial p_{s,d}}{\partial z} - \beta_{dc}\left(\frac{U_{s,c}}{1-\varepsilon_c} - \frac{U_{s,d}}{1-\varepsilon_d}\right) \quad \text{(T3.1)}$$

(2) One-dimensional transient momentum equation of clusters

$$\frac{\partial}{\partial t}(f^2\rho_s U_{s,c}) + \frac{\partial}{\partial z}\left[(f^2\rho_s U_{s,c})\frac{fU_{s,c}}{(1-\varepsilon_c)}\right] = -f(1-\varepsilon_c)\frac{\partial p}{\partial z} + n_c F_c + n_i F_i - \frac{\partial p_{s,c}}{\partial z} - f(1-\varepsilon_c)\rho_s g + \beta_{dc}\left(\frac{U_{s,d}}{1-\varepsilon_d} - \frac{U_{s,c}}{1-\varepsilon_c}\right) \quad \text{(T3.2)}$$

(3) One-dimensional transient momentum equations of gas in the dilute and dense phases

$$\frac{\partial}{\partial t}[(1-f)^2\rho_g U_{g,d}] + \frac{\partial}{\partial z}\left[(1-f)^2\rho_g U_{g,d}\frac{(1-f)U_{g,d}}{\varepsilon_d}\right] = -(1-f)\varepsilon_d\frac{\partial p}{\partial z} - n_d F_d - n_i F_i - (1-f)\varepsilon_d\rho_g g \quad \text{(T3.3)}$$

$$\frac{\partial}{\partial t}(f^2\rho_g U_{g,c}) + \frac{\partial}{\partial z}\left[(f^2\rho_g U_{g,c})\frac{fU_{g,c}}{\varepsilon_c}\right] = -f\varepsilon_c\frac{\partial p}{\partial z} - n_c F_c - f\varepsilon_c\rho_g g \quad \text{(T3.4)}$$

(continued)

**Table 3.1** (continued)

(4) One-dimensionalkinetic energy equations of particlesand gas

$$\frac{\partial E_s}{\partial t}+\frac{\partial}{\partial z}[f^2\rho_s E_{s,c}U_{s,c}+(1-f)^2\rho_s E_{s,d}U_{s,d}]+\alpha_s v_s\frac{\partial p}{\partial z}+$$
$$\frac{\partial}{\partial z}\left[\left(\frac{fU_{s,c}}{1-\varepsilon_c}\right)p_{s,c}+\left(\frac{(1-f)U_{s,d}}{1-\varepsilon_d}\right)p_{s,d}\right]+\frac{\partial}{\partial z}(q_{s,d}+q_{s,c})+[f^2\rho_s U_{s,c}+(1-f)^2\rho_s U_{s,d}]g \quad \text{(T3.5)}$$
$$=[n_d F_d\frac{(1-f)U_{s,d}}{1-\varepsilon_d})+n_c F_c\frac{fU_{s,c}}{1-\varepsilon_c}+n_i(1-f)F_i U_i]+(\Gamma_{s,d}+\Gamma_{s,c})$$

$$\frac{\partial E_g}{\partial t}+\frac{\partial}{\partial z}[f^2\rho_g E_{g,c}U_{g,c}+(1-f)^2\rho_g E_{g,d}U_{g,d}]$$
$$+\alpha_g v_g\frac{\partial p}{\partial z}+\frac{\partial}{\partial z}(q_{g,d}+q_{g,c})+[f\rho_g U_{g,c}+(1-f)\rho_g U_{g,d}]g \quad \text{(T3.6)}$$
$$=-\left[n_d F_d\frac{(1-f)U_{g,d}}{\varepsilon_d}+n_c F_c\frac{fU_{g,c}}{\varepsilon_c}\right]+(\Gamma_{g,d}+\Gamma_{g,c})$$

(5) Drag forces of gas and particles in the dilute and dense phases

$$F_d=\frac{\pi d_p^2\rho_g}{8\varepsilon_d^{2.65}}\left[\frac{24}{Re_d}+\frac{3.6}{Re_d^{0.313}}\right]\left|\frac{U_{g,d}}{\varepsilon_d}-\frac{U_{s,d}}{1-\varepsilon_d}\right|\left(\frac{U_{g,d}}{\varepsilon_d}-\frac{U_{s,d}}{1-\varepsilon_d}\right) \quad \text{(T3.7)}$$

$$F_c=\frac{\pi d_p^2\rho_g}{8\varepsilon_c^{2.65}}\left[\frac{24}{Re_c}+\frac{3.6}{Re_c^{0.313}}\right]\left|\frac{U_{g,c}}{\varepsilon_c}-\frac{U_{s,c}}{1-\varepsilon_c}\right|\left(\frac{U_{g,c}}{\varepsilon_c}-\frac{U_{s,c}}{1-\varepsilon_c}\right) \quad \text{(T3.8)}$$

$$F_i=\frac{\pi d_c^2\rho_g\varepsilon_d^2}{8(1-f)^{2.65}}\left[\frac{24}{Re_i}+\frac{3.6}{Re_i^{0.313}}\right]\left|\left(\frac{U_{g,d}}{\varepsilon_d}-\frac{U_{s,c}}{1-\varepsilon_c}\right)(1-f)\right|\left[\left(\frac{U_{g,d}}{\varepsilon_d}-\frac{U_{s,c}}{1-\varepsilon_c}\right)(1-f)\right] \quad \text{(T3.9)}$$

(continued)

**Table 3.1** (continued)

| |
|---|
| (6) Drag coefficient of particles between the dilute phase and the dense phase |
| $\beta_{dc} = \frac{3(1+e)\rho_s(1-\varepsilon_d)(1-\varepsilon_c)}{d_p}\left[\frac{1}{\alpha_g} + \frac{3(2-\varepsilon_d-\varepsilon_c)}{2\alpha_g^2}\right]\left\|\frac{U_{s,c}}{1-\varepsilon_c} - \frac{U_{s,d}}{1-\varepsilon_d}\right\|$ (T3.10) |
| (7) Solids pressure of the dilute phase and dense phase |
| $p_{s,d} = \rho_s(1-f)(1-\varepsilon_d)\left[1 + 2g_{0,d}(1-f)(1-\varepsilon_d)(1+e)\right]\theta_d$ (T3.11) |
| $p_{s,c} = \rho_s f(1-\varepsilon_c)\left[1 + 2g_{0,c} f(1-\varepsilon_c)(1+e)\right]\theta_c$ (T3.12) |
| (8) Number density of cluster, dispersed particles and interface |
| $n_c = \frac{f(1-\varepsilon_c)}{\pi d_p^3/6}, n_d = \frac{(1-f)(1-\varepsilon_d)}{\pi d_p^3/6}, n_i = \frac{f}{\pi d_c^3/6}$ (T3.13) |
| (9) Radial distribution function of dilute and dense phases |
| $g_{0,d} = \frac{\alpha_{s,\max}^{1/3}}{\{\alpha_{s,\max}^{1/3} - [(1-f)(1-\varepsilon_d)]^{1/3}\}}, g_{0,c} = \frac{\alpha_{s,\max}^{1/3}}{\{\alpha_{s,\max}^{1/3} - [f(1-\varepsilon_c)]^{1/3}\}}$ (T3.14) |
| (10) Collisional energy dissipations of dispersed particles and clusters |
| $\Gamma_{s,d} = \frac{3(1-e^2)[(1-f)(1-\varepsilon_d)]^2\rho_s\alpha_{s,max}^{1/3}\theta_d}{\alpha_{s,max}^{1/3} - [(1-f)(1-\varepsilon_d)]^{1/3}}\left[\frac{4}{d_p}\sqrt{\frac{\theta_d}{\pi}} - \frac{\partial}{\partial z}\frac{(1-f)U_{s,d}}{1-\varepsilon_d}\right]$ (T3.15) |
| $\Gamma_{s,c} = \frac{3(1-e^2)[f(1-\varepsilon_c)]^2\rho_s\alpha_{s,max}^{1/3}\theta_c}{\{\alpha_{s,max}^{1/3} - [f(1-\varepsilon_c)]^{1/3}\}}\left[\frac{4}{d_p}\sqrt{\frac{\theta_c}{\pi}} - \frac{\partial}{\partial z}\frac{fU_{s,c}}{1-\varepsilon_c}\right]$ (T3.16) |
| (11) Rate of energy dissipations of gas in the dilute and dense phases |
| $\Gamma_{g,d} = \frac{4}{3}(1-f)\varepsilon_d\mu_g\left[\frac{\partial}{\partial z}\left(\frac{(1-f)U_{g,d}}{\varepsilon_d}\right)\right]^2$ (T3.17) |

(continued)

**Table 3.1** (continued)

| |
|---|
| $\Gamma_{g,c} = \frac{4}{3} f \varepsilon_c \mu_g \left[ \frac{\partial}{\partial z} \left( \frac{f U_{g,c}}{\varepsilon_c} \right) \right]^2$ (T3.18) |
| (12) Coefficients |
| $K_{1,d} = 2(1+e)\rho_s g_{0,d}$, $K_{1,c} = 2(1+e)\rho_s g_{0,c}$ (T3.19) |
| $K_{2,c} = \frac{8 d_p \rho_s (1+e) f (1-\varepsilon_c) g_{0,c}}{(3\sqrt{\pi})} - \frac{d_p \rho_s}{6} \left\{ \frac{\sqrt{\pi}}{3(3-e)} \left[1 + 0.4(1+e)(3e-1) f (1-\varepsilon_c) g_{0,c}\right] + \frac{8 f (1-\varepsilon_c) g_{0,c} (1+e)}{5\sqrt{\pi}} \right\}$ (T3.20) |
| $K_{2,d} = \frac{8 d_p \rho_s (1+e)(1-f)(1-\varepsilon_d) g_{0,d}}{(3\sqrt{\pi})} - \frac{d_p \rho_s}{6} \left\{ \frac{\sqrt{\pi}}{3(3-e)} \left[1 + 0.4(1+e)(3e-1)(1-f)(1-\varepsilon_d) g_{0,d}\right] + \frac{8(1-f)(1-\varepsilon_d) g_{0,d} (1+e)}{5\sqrt{\pi}} \right\}$ (T3.21) |
| $K_{3,c} = \frac{d_p \rho_s}{4} \left\{ \frac{\sqrt{\pi}\left[1+0.4(1+e)(3e-1) f (1-\varepsilon_c) g_{0,c}\right]}{3(3-e)} + \frac{8 f (1-\varepsilon_c) g_{0,c} (1+e)}{5\sqrt{\pi}} \right\}$ (T3.22) |
| $K_{3,d} = \frac{d_p \rho_s}{4} \left\{ \frac{\sqrt{\pi}\left[1+0.4(1+e)(3e-1)(1-f)(1-\varepsilon_d) g_{0,d}\right]}{3(3-e)} + \frac{8(1-f)(1-\varepsilon_d) g_{0,d} (1+e)}{5\sqrt{\pi}} \right\}$ (T3.23) |
| $K_{4,c} = \frac{12\left(1-e^2\right)\rho_s g_{0,c}}{d_p \sqrt{\pi}}$, $K_{4,d} = \frac{12\left(1-e^2\right)\rho_s g_{0,d}}{d_p \sqrt{\pi}}$ (T3.24) |
| $K_{5,c} = \frac{12\left(1-e^2\right) f (1-\varepsilon_c) \rho_s g_{0,c}}{d_p \sqrt{\pi}}$, $K_{5,d} = \frac{12\left(1-e^2\right)(1-f)(1-\varepsilon_d) \rho_s g_{0,d}}{d_p \sqrt{\pi}}$ (T3.25) |
| $K_{6,c} = \left[ \frac{96\left(1-e^2\right) f (1-\varepsilon_c) \rho_s g_{0,c}}{d_p \sqrt{\pi}} \right]^2$, $K_{6,d} = \left[ \frac{96\left(1-e^2\right)(1-f)(1-\varepsilon_d) \rho_s g_{0,d}}{d_p \sqrt{\pi}} \right]^2$ (T3.26) |

respect to time and the spatial gradients of velocity of dispersed particles in the dilute phase. $\beta_{dc}$ is the drag coefficient of the solids-solids momentum transfer between the dilute phase and the dense phase.

We assume that the off-diagonal elements of the gradient of the solid stress tensor of clusters are zero in a first approximation. The one-dimensional transient momentum equation of clusters is expressed by Eq. (T3.2), as shown in Table 3.1 (Huilin 2017; Dan et al. 2018). The second and third terms on the right hand side represent the drag forces in the dense phase and the interface between the dilute phase and the dense phase. Because the gas density is less than that of particles, Eq. (T3.2) can be rewritten to

$$n_c F_c + n_i F_i = f(1-\varepsilon_c)(\rho_s - \rho_g)(g + a_{sc,l} + a_{sc,c}) + f(1-\varepsilon_c)\frac{\partial p}{\partial z} + \frac{\partial p_{s,c}}{\partial z} - \beta_{dc}\left(\frac{U_{s,d}}{1-\varepsilon_d} - \frac{U_{s,c}}{1-\varepsilon_c}\right) \quad (3.51)$$

where $a_{sc,l}$ and $a_{sc,c}$ are the cluster temporal acceleration and convective acceleration of the dense phase, and represent the rate of cluster velocity change with respect to time and the spatial gradients of velocity of clusters due to the change of position. Both accelerations can be deduced from the given equations.

Similar to the assumptions used in the momentum equations of dispersed particles and clusters, the one-dimensional transient momentum equations of gas phase in the dilute phase and the dense phase along the direction of the flow are expressed by Eqs. (T3.3) and (T3.4) (Huilin 2017; Dan et al. 2018). Combining with them, the gas momentum equation becomes as follows:

$$\frac{n_c F_c}{f\varepsilon_c} = \frac{n_d F_d}{(1-f)\varepsilon_d} + \frac{n_i F_i}{(1-f)\varepsilon_d} + \rho_g[(a_{gd,c} - a_{gc,c}) + (a_{gd,l} - a_{gc,l})] \quad (3.52)$$

where the gas temporal accelerations $a_{gc,l}$ and $a_{gd,l}$, and gas convective accelerations $a_{gc,c}$ and $a_{gd,c}$ of the dense phase and dilute phase can be deduced from thegiven equations.

Mass conservation of gas and particles includes mass balance in the dilute and dense phases. The gas and particles mass balance equations in the dense phase and dilute phase are expressed in the computational cell as following.

$$f\rho_g U_{g,c} + (1-f)\rho_g U_{g,d} = \alpha_g \rho_g v_g \quad (3.53)$$

$$f\rho_s U_{s,c} + (1-f)\rho_s U_{s,d} = (1-\alpha_g)\rho_s v_s \quad (3.54)$$

The gas volume fraction requires that the sum of volume fraction of gas and particles be unity in the computational cell, that is,

$$\alpha_g = f\varepsilon_c + (1-f)\varepsilon_d \tag{3.55}$$

The fluctuating energy equations describe granular temperature equations for dispersed particles and clusters. The algebraic expression for granular temperature proposed by Syamlal et al. (1993) assumes that the granular energy is dissipated locally; neglecting the convection and diffusion contributions, and retaining the generation and dissipation terms. In present study, the algebraic expression model is extended to predict granular temperatures of dispersed particles $\theta_d$ and granular temperature of particles within the clusters $\theta_c$ in a first approximation.

$$\theta_d = \left\{\frac{-K_{1,d}}{2K_{4,d}}\frac{\partial}{\partial z}\left[\frac{(1-f)U_{s,d}}{(1-\varepsilon_d)}\right] + \sqrt{\left(\frac{K_{1,d}^2}{4K_{4,d}^2} + \frac{K_{2,d}}{K_{5,d}}\right)\frac{\partial^2}{\partial z^2}\left[\frac{(1-f)U_{s,d}}{(1-\varepsilon_d)}\right] + \frac{2K_{3,d}}{K_{6,d}}\frac{\partial}{\partial z}\left[\frac{(1-f)U_{s,d}}{(1-\varepsilon_d)}\right]^2}\right\}^2 \tag{3.56}$$

$$\theta_c = \left\{\frac{-K_{1,c}}{2K_{4,c}}\frac{\partial}{\partial z}\left[\frac{fU_{s,c}}{(1-\varepsilon_c)}\right] + \sqrt{\left(\frac{K_{1,c}^2}{4K_{4,c}^2} + \frac{K_{2,c}}{K_{5,c}}\right)\frac{\partial^2}{\partial z^2}\left[\frac{fU_{s,c}}{(1-\varepsilon_c)}\right] + \frac{2K_{3,c}}{K_{6,c}}\frac{\partial}{\partial z}\left[\frac{fU_{s,c}}{(1-\varepsilon_c)}\right]^2}\right\}^2 \tag{3.57}$$

where the coefficients $K_i$ ($i$ = 1–6) are given in Table 3.1. $e$ is the coefficient of restitution of particles.

### 3.6.5 *Conditioned Extreme Value Equation Based on BEV Theory*

It is important to keep in mind that we solve an extreme value Eq. (3.44) of the minimization of energy dissipation rate, instead of solving the balance Eq. (3.40) of granular energy $E_s$ and gas energy $E_g$ to determine the heterogeneous structure parameters of the dense phase and dilute phase in CVs. Like energy balance Eq. (3.40), the extreme value Eq. (3.44) takes as a constitutive equation for flow of heterogeneous structures.

For flow of heterogeneous structures with dispersed particles in the dilute phase and clusters in the dense phase, there are ten independent variables at the state I, as shown in Fig. 3.2. On the other hand, we have a set of equations with an extreme value Eq. (3.44) and eight balance equations including three momentum Eqs. (3.50)–(3.52), three continuity Eqs. (3.53)–(3.55) and two granular temperature Eqs. (3.56)–(3.57). Hence, the set of transient nonlinear equations is unclosed to solve ten heterogeneous structure parameters in CVs.

The dependence function with two independent variables is described by the bivariate extreme value (BEV) theory (Kotz and Nadarajah 2000). The BEV method determines an extreme point of the dependence function with the change of two independent variables if it has a solution. Thus, the values of two independent variables are determined from the minimum value point of the dependence function. In present study, among ten heterogeneous structure parameters the gas volume fractions of the dense phase $\varepsilon_c$ and the dilute phase $\varepsilon_d$ are chosen as two independent variables of the dependence function $N_{df}$. Using BEV theory, the dependence function $N_{df}$ from Eq. (3.44) is expressed as follows

$$N_{df}^{BEV} = (\Omega_{dr} + \Omega_{co} + \Omega_{vi})|_{\varepsilon_d \in R_1, \varepsilon_c \in R_2}^{\alpha_g, p, v_g, v_s, \theta} = \min \tag{3.58}$$

This is a conditioned extreme value equation, and it gives the relation between the minimum value point of the dependence function $N_{df}$ and two heterogeneous structure parameters $\varepsilon_d$ and $\varepsilon_c$ in the ranges of the dilute phase $R_1$ and the dense phase $R_2$ according to five CV parameters in the CV. The ranges of $R_1$ and $R_2$ are $R_1 \in [\alpha_g, 1.0]$ and $R_2 \in [\alpha_{mf}, \alpha_g]$, where $\alpha_{mf}$ is the gas volume fraction at the minimum fluidization. The conditioned extreme value Eq. (3.54) indicates that the dependence function $N_{df}$ has a minimum value point in the range of $R_1$ and $R_2$ if the heterogeneous structure exists in the CV. Thus, the heterogeneous structure parameters $\varepsilon_d$ and $\varepsilon_c$ of the computational cell are determined at the minimum value point. Furthermore, once the existence of clusters is identified in the CV, the remaining eight heterogeneous structure parameters (i.e., $d_c$, $f$, $U_{g,d}$, $U_{s,d}$, $\theta_d$, $U_{g,c}$, $U_{s,c}$ and $\theta_c$) are obtained by solving eight balance equations. Hence, the set of transient nonlinear equations is determined.

On the other hand, Eq. (44) has no extreme point, and the rate of energy dissipation $\Omega$ is monotonically increased with the change of gas volume fractions of the dilute phase $\varepsilon_d$ in the range $R_1$ and the dense phase $\varepsilon_c$ in the range $R_2$. This identifies a homogenous flow (HF) of gas and particles mixture at the state II, as shown in Fig. 3.2 in the CV.

$$N_{df}^{BEV} = (\Omega_{dr} + \Omega_{co} + \Omega_{vi})|_{\varepsilon_d \in R_1, \varepsilon_c \in R_2}^{\alpha_g, p, v_g, v_s, \theta} \neq \min \tag{3.59}$$

The solutions of Eqs. (3.58) and (3.59) determine the heterogeneity of clusters and the homogeneity of particles suspension in the CVs. They indicate that the heterogeneous structures exist when the dependence function of energy dissipation rate has a minimum value point. The existence of the minimum value point of the energy dissipation rate depends on the CV parameters, suggesting that the direction of evolution of heterogeneous structure minimizes its energy dissipation rate compatible with the constrains of CV parameters.

### 3.6.6 Minimization of Energy Dissipation Rate of Heterogeneous Structure

The riser height and diameter are 6.58 m and 7.5 cm. The average diameter and density of FCC particles are 75 μm and 1650 kg/m$^3$, respectively. The two dimensional computational domain is used in the numerical simulations. The exit gas pressure is assumed to be 101,330 N/m$^2$. The particles-wall restitution coefficient is estimated from the experimental measurements (Gidaspow and Huilin 1998) in the riser.

The criterion of cluster existence indicates that the energy dissipation rate is minimum as the gas volume fractions of $\varepsilon_d$ and $\varepsilon_c$ vary in the ranges of $R_1$ for the dilute phase and $R_2$ for the dense phase. The profile of instantaneous energy dissipation rate is shown in Fig. 3.3 at two different CVs. Both the CV instantaneous gas velocity and solid velocity are positive, indicating a co-current up-flow of gas and particles. In the CV (a), several extreme points of energy dissipation rates are found with the change of $\varepsilon_c$ and $\varepsilon_d$. The multi-extreme point behavior originates from the contributions of hydrodynamic interactions and collisional collisions of particles. The minimum value point is found from these extreme points, indicating the clusters exist in the CV. The values of $\varepsilon_d$ and $\varepsilon_c$ at the minimum value point are 0.982 and 0.840. The remaining eight heterogeneous structure parameters are calculated by means of the equation set, and are listed in Table 3.2. On the other hand, for the CV (b) the energy dissipation rate is monotonically increased with the increase of $\varepsilon_c$ and decrease of $\varepsilon_d$. There are no extreme points with the change of $\varepsilon_d$ in the range $R_1$ and $\varepsilon_c$ in the range $R_2$, indicating that the HF of gas-particle suspension exists in the CV.

Figure 3.4 shows the distributions of instantaneous energy dissipation rate at two different CVs. Both cases have a positive gas velocity and negative solid velocity, indicating a counter-current flow with gas up-flow and particles down-flow in the CVs. In the CV (a), the distribution of energy dissipation rate has several extreme points with the change of $\varepsilon_c$ and $\varepsilon_d$. One minimum value point is determined, and the values of gas volume fractions of the dilute phase and dense phase are 0.976 and 0.799 at the minimum value point. The gas superficial velocities of the dilute phase and dense phase shown in Table 3.2 are positive. The dispersed particles superficial velocity of the dilute phase $U_{s,d}$ is positive and the cluster superficial velocity of the dense phase $U_{s,c}$ is negative, suggesting the clusters flow downward and the dispersed particles flow upward in the CV. On the other hand, the energy dissipation

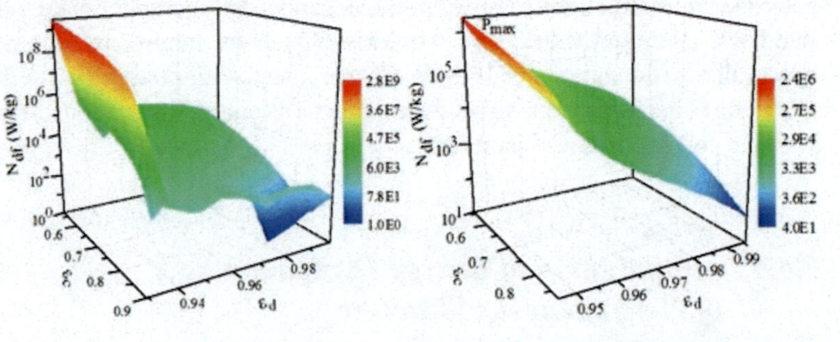


(a)$u_g$=3.01m/s, $u_s$=2.27m/s, $\alpha_s$=0.072, $x/D$=0.089 (b)$u_g$=0.881m/s, $u_s$=0.815m/s, $\alpha_s$=0.058, $x/D$=0.45

$\Delta p/\Delta z$=-4863.2 Pa/m, $\theta$=2.11x10$^{-3}$ (m/s)$^2$, $H$=4.56m $\Delta p/\Delta z$=-2430.3 Pa/m, $\theta$=1.61x10$^{-3}$ (m/s)$^2$, $H$=4.93m

**Fig. 3.3** Instantaneous rates of energy dissipations of gas-particles co-current upflow in two CVs

**Table 3.2** Heterogeneous structure parameters of the minimum value point at CV parameters

| Parameters | | | Co-current up-flow | Co-current down-flow | Counter-current flow |
|---|---|---|---|---|---|
| CV parameters | | $v_g$(m/s) | 3.01 | −1.52 | 0.055 |
| | | $v_s$(m/s) | 2.27 | −1.58 | −0.072 |
| | | $\alpha_g$ | 0.928 | 0.879 | 0.962 |
| | | $\theta$ (m/s) $^2$ | $2.11 \times 10^{-3}$ | $1.23 \times 10^{-3}$ | $2.75 \times 10^{-3}$ |
| | | $\Delta p/\Delta z$(Pa/m) | −4863.2 | −2431.1 | −2400.4 |
| Heterogeneous structure parameters | At minimum value point | $\varepsilon_d$ | 0.982 | 0.976 | 0.983 |
| | | $\varepsilon_c$ | 0.840 | 0.799 | 0.866 |
| | Equation set | $f$ | 0.377 | 0.547 | 0.177 |
| | | $d_c$ (m) | $4.29 \times 10^{-3}$ | $7.92 \times 10^{-4}$ | $5.45 \times 10^{-4}$ |
| | | $U_{s,d}$ (m/s) | $6.11 \times 10^{-2}$ | $-3.01 \times 10^{-3}$ | $-2.56 \times 10^{-3}$ |
| | | $U_{s,c}$ (m/s) | 0.331 | −0.324 | $-3.69 \times 10^{-3}$ |
| | | $U_{g,d}$ (m/s) | 3.46 | −1.413 | $6.39 \times 10^{-2}$ |
| | | $U_{g,c}$ (m/s) | 1.68 | −1.28 | $3.20 \times 10^{-3}$ |
| | | $\theta_d$ (m/s)$^2$ | $4.60 \times 10^{-3}$ | $3.66 \times 10^{-2}$ | $1.99 \times 10^{-3}$ |
| | | $\theta_c$ (m/s)$^2$ | $6.21 \times 10^{-5}$ | $0.10 \times 10^{-4}$ | $0.25 \times 10^{-5}$ |

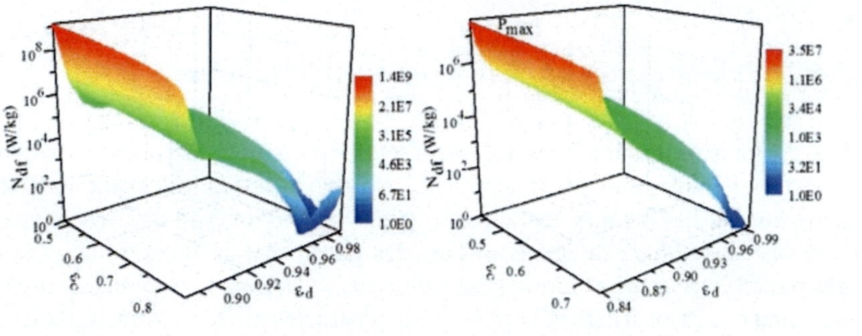


(a)$v_g$=-1.52m/s, $v_s$=-1.58m/s, $H$=4.77m, $x/D$=0.19 (b)$v_g$=-1.32m/s, $v_s$=-1.40m/s, $H$=3.57m, $x/D$=0.38

$\alpha_s$=0.121, $\Delta p/\Delta z$=-2431.1 Pa/m, $\theta$=1.23x10$^{-3}$ (m/s)$^2$ $\alpha_s$=0.165, $\Delta p/\Delta z$=-3125.3 Pa/m, $\theta$=5.77x10$^{-3}$ (m/s)$^2$

**Fig. 3.4** Instantaneous rates of energy dissipations of co-current downward flow in two CVs

rate in the CV (b) increases with the increase of $\varepsilon_c$ and $\varepsilon_d$, indicating the HF of gas and particles suspension in the CV.

The profiles of instantaneous energy dissipation rate are shown in Fig. 3.5 at two CVs. The CV parameters of gas velocity and solid velocity are negative, indicating a co-current down-flow of gas and particles. In the CV (a), the heterogeneous flow

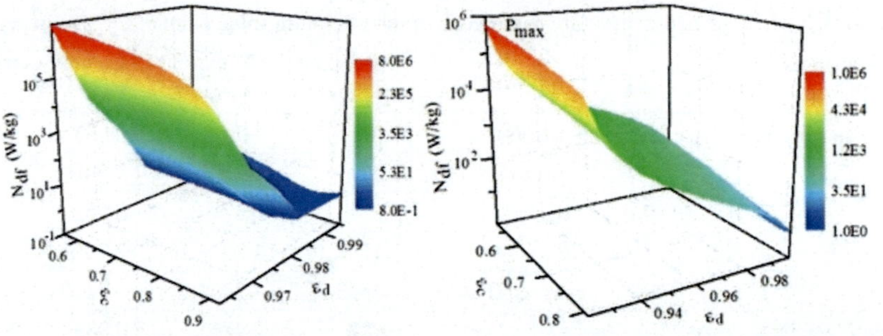


(a)$v_g$=0.055m/s, $v_s$=-0.072m/s, $H$=3.61m, $x/D$=0.19 (b)$v_g$=0.023m/s, $v_s$=-0.197m/s, $H$=5.05m, $x/D$=0.38

$\alpha_s$=0.038, $\Delta p/\Delta z$=-2400.4 Pa/m, $\theta$=2.75x10$^{-3}$ (m/s)$^2$ $\alpha_s$=0.082, $\Delta p/\Delta z$=-2410.3 Pa/m, $\theta$=3.15x10$^{-2}$ (m/s)$^2$

**Fig. 3.5** Instantaneous rates of energy dissipations of gas upflow and particles downflowin two CVs

is identified from the extreme points of energy dissipation rate. The gas volume fractions of the dense phase and dilute phase are 0.983 and 0.866 at the minimum value point of energy dissipation rate. The superficial velocities of gas, dispersed particles and clusters in the dense phase and dilute phase, as shown in Table 3.2, are negative. On the other hand, the HF in the CV (b) is identified because no extreme points exist with the change of $\varepsilon_d$ in the range $R_1$ and $\varepsilon_c$ in the range $R_2$ in the CV.

### 3.6.7 Intermittency of Heterogeneous Structures

The contours of instantaneous volume fractions of particles and energy dissipation rates at the minimum value point $N_{df,min}$ are shown in Fig. 3.6 at the inlet solids mass flux of 20.4 kg/m$^2$s and superficial gas velocity of 3.48 m/s. The simulated instantaneous volume fractions of particles reveal that an initially homogeneous gas-particles suspension forms gradually a heterogeneous structure. The opposite direction from a local high volume fraction of particles to a local low volume fraction is also found in the different CVs. These variations of volume fraction of particles represent the formation of heterogeneous structures in the CVs, indicating that the flow structures of particles depend on energy dissipation rate. When the CV has a minimum value point $N_{df,min}$, it is safe to say that the cluster exists in the CV. On the contrary, the white color in Fig. 3.6 represents homogeneous flow of gas-particles suspension in the CVs. It indicates that the $N_{df,min}$ reveals the existence of heterogeneous structures in the CVs.

The instantaneous gas velocity $v_g$, solids volume fraction $\alpha_s$ and gas volume fractions of $\varepsilon_c$ and $\varepsilon_d$ at the minimum value point of energy dissipation rate $N_{df,min}$ are shown in Fig. 3.7 in the CV. The denser clusters with low $\varepsilon_c$ are formed near the wall, and the diluter clusters with high $\varepsilon_c$ exist at the center of the riser. From simulated

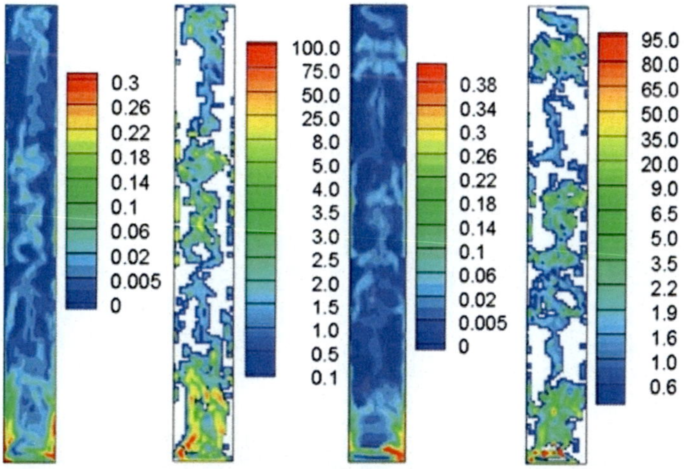


(a) $t$=31s: $\alpha_s$ $N_{df,min}$ (b) $t$=35s: $\alpha_s$ $N_{df,min}$

**Fig. 3.6** Instantaneous volume fractions of particles and minimum energy dissipation rates (The color of $N_{df,min}$ represents clusters flow, and the white color is homogeneous flow without clusters in CVs)

**Fig. 3.7** Profiles of computational cell parameters and heterogeneous structure parameters

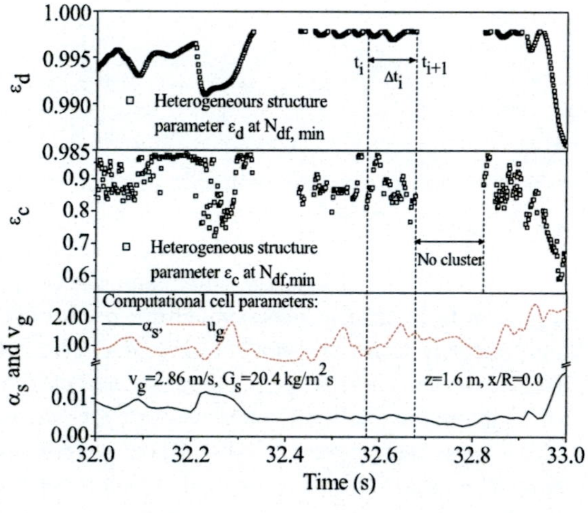


instantaneous gas volume fractions of $\varepsilon_d$ and $\varepsilon_c$, the occurrence of heterogeneous structures is confined within a more clearly defined boundary from the starting time $t_i$ to the ending time $t_{i+1}$. The duration time of clusters $\Delta t_i$ reveals the cluster existence time in the CV. The variation of the duration time of clusters proves intermittency between the heterogeneous flow and the homogeneous flow in the CV.

For the single fluid transition flow, the intermittency is given to the alternation between turbulent and nonturbulent motion that characterizes certain types of fluid

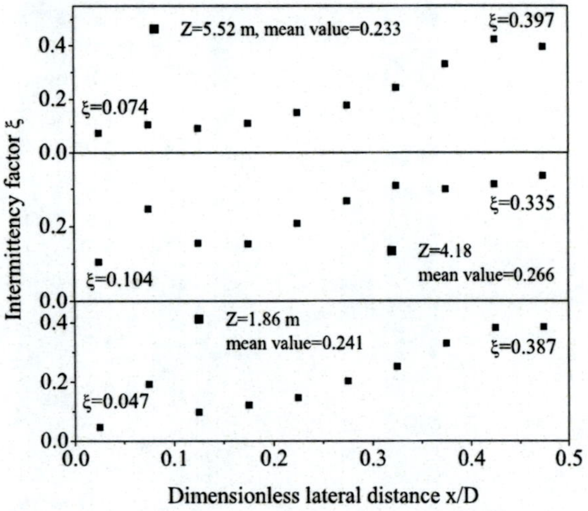


**Fig. 3.8** Distribution of intermittency factor at three riser heights

flow (Libby 1975). The intermittency factor is defined, and it measures the probability of finding turbulent flow at any instant of the point considered (Good et al. 2012). Similarly, the intermittency factor of the flow of heterogeneous structures of gas-particles mixture is defined as

$$\xi = \frac{\sum_{i=1} \Delta\tau_i}{\tau} \tag{3.60}$$

where $\tau$ is the total sampling time in the CV. The intermittency factor measures the existence time fraction of heterogeneous structures in the CV. It can be seen that the intermittency factor of 0.0 characterizes a HF of gas-particles suspension and no clusters exit in the CV. On the other hand, the intermittency factor of 1.0 represents a flow of fully heterogeneous structures of clusters in the CV. The distributions of intermittency factor are shown in Fig. 3.8 at three riser heights. The intermittency factor is lower at the riser center than that near the riser wall, indicating that the values of the intermittency factor increase toward the riser walls. The values of intermittency factor are in the range of 0.047–0.104 at the riser center, and 0.335–0.397 near the wall. The mean value of intermittency factor is from 0.233 to 0.266, indicating that the flow of heterogeneous structures is intermittent in the CVs.

### *3.6.8 Compromise of Energy Dissipations Between Gas-Particles and Inter-Particle Interactions*

The formation of heterogeneous structure contributes to the interactions of particles with other particles and gas phase. The inelastic collisions of particles dissipate kinetic energy $\Omega_{co}$, and the hydrodynamic interactions between the gas phase and solids phase results in drag energy dissipation $\Omega_{dr}$. The gas shear strain rates in the dense and dilute phases give rise to gas viscous energy dissipation $\Omega_{vi}$. Figure 3.9 shows the profile of instantaneous energy dissipation components at the minimum point. The simulated drag energy dissipation component $\Omega_{dr}$ and collisional energy dissipation component $\Omega_{co}$ are of the same order of magnitude. However, the gas viscous energy dissipation component $\Omega_{vi}$ is several orders of magnitude smaller than the drag and collisional energy dissipation components. Therefore, the gas viscous energy dissipation component $\Omega_{vi}$ is negligible in the determination of the minimum value point of energy dissipation rate $N_{df}$ of the CVs.

As the order of magnitude of gas viscous energy dissipation componentis much less than drag energy dissipation and collisional energy dissipation components, Eq. (3.58) can further be simplified to the following equation

$$N_{df}^{BEV} \approx (\Omega_{dr} + \Omega_{co})|_{\varepsilon_d \in R_1, \varepsilon_c \in R_2}^{\alpha_g, p, v_g, v_s, \theta} = \min \tag{3.61}$$

The instantaneous energy dissipation components $\Omega_{dr,\min}$ and $\Omega_{co,\min}$ are shown in Fig. 3.10. The fluctuations of energy dissipation components contribute to the drag forces and the interactions of collisions, leading to the variation of energy dissipation components with time. The origin of clusters traces to the dissipative nature of gas-particles interactions $\Omega_{dr}$ and particle–particle collisions $\Omega_{co}$. The onset and the cessation of particle clustering have been detected from instantaneous energy dissipation rate components. The life (i.e. duration) of the heterogeneous structure defines the time in the range from the onset (ON) of particles clustering to the cessation (OFF) of clusters. During the period of the duration of the cluster, the heterogeneous structure flow is controlled by $\Omega_{co,min}$ at the onset, and then $\Omega_{dr,min}$ is dominant. Finally, the $\Omega_{co,min}$ results in the cessation of the cluster, indicating the formation and breaking of heterogeneous structures relate with $\Omega_{dr,min}$ and $\Omega_{co,min}$

**Fig. 3.9** Profiles of energy dissipation components of heterogeneous structures

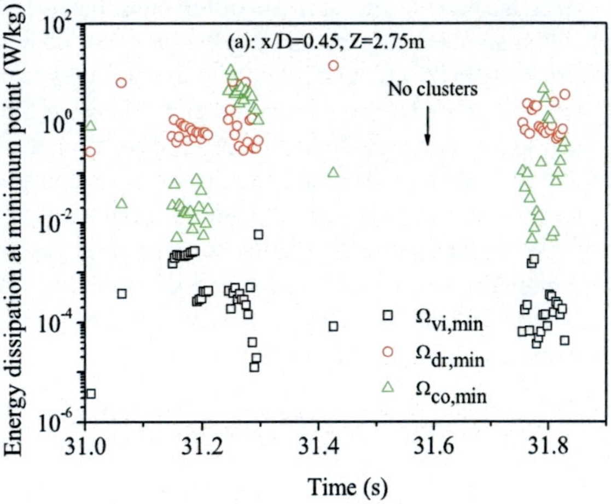

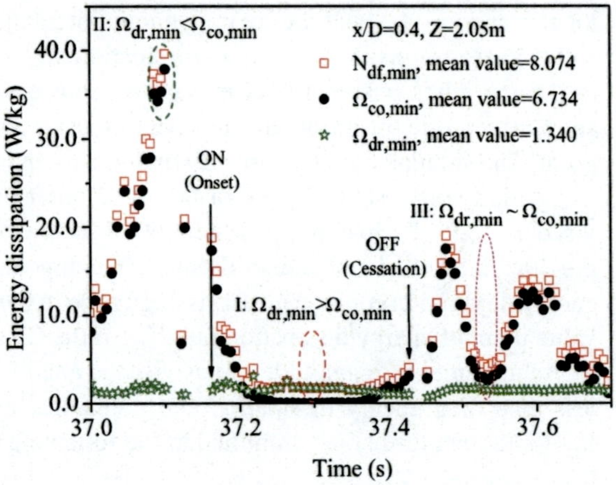


**Fig. 3.10** Instantaneous rates of energy dissipations at two different CVs

in the CV. The instantaneous $\Omega_{dr,min}$ is larger than $\Omega_{co,min}$ at time I. However, it is opposite at time II, and the instantaneous $\Omega_{co,min}$ is larger than $\Omega_{dr,min}$. On the other hand, the instantaneous $\Omega_{dr,min}$ is close to $\Omega_{co,min}$ at time III, indicating that the different dominant mechanisms of the formation of heterogeneous structures exist in the center and near the riser walls.

When the energy dissipation component of gas-particles drag forces is smaller than the energy dissipation component of collisional interactions, the conditioned extreme value Eq. (3.61) simplifies to

$$N_{df}^{BEV} \approx N_{df,CD}^{BEV} = \Omega_{co}\Big|_{\varepsilon_d \in R_1, \varepsilon_c \in R_2}^{\alpha_g, p, v_g, v_s, \theta} = \min \qquad \Omega_{dr} < \Omega_{co} \tag{3.62}$$

The minimization of energy dissipation rate is equivalent to the minimization of energy dissipation component of collisional interactions of heterogeneous structures. It appears that $\Omega_{dr} < \Omega_{co}$ must be satisfied for Eq. (3.61) to be a valid approximation since the error in Eq. (3.62) is of the order $O(\Omega_{dr})$. This indicates the interactions of collisions are dominant, and it is known as a collision-dominant (CD) heterogeneous structure. In the granular homogenous cooling system (HCS) with high solids volume fractions, the particles are agglomerated because the granular temperature decreases due to particle–particle collisions (Haff et al. 1983; McMillan et al. 2013; Yin et al. 2013). The formation of heterogeneous structure could arise when the collisions between the particles are sufficiently inelastic.

When the energy dissipation component of gas-particles drag forces is larger than the energy dissipation component of collisional interactions, the energy dissipation component of hydrodynamic interactions is dominant. The conditioned extreme value Eq. (3.61) simplifies to

$$N_{df}^{BEV} \approx N_{df,HD}^{BEV} = \Omega_{dr}\Big|_{\varepsilon_d \in R_1, \varepsilon_c \in R_2}^{\alpha_g, p, v_g, v_s, \theta} = \min \qquad \Omega_{dr} > \Omega_{co} \tag{3.63}$$

It appears that $\Omega_{dr} > \Omega_{co}$ is satisfied for Eq. (3.61) to be a valid approximation since the error in Eq. (3.63) is of the order $0(\Omega_{co})$, and it is known as a hydrodynamic-dominant (HD) heterogeneous structure. The formation of clusters arises from the hydrodynamic forces of heterogeneous structures. Wylie and Koch (2000) found that the formation of clusters was driven by the energy dissipation due to interactions surrounding gas and particles. The dilute clusters with low solids volume fraction increase slip velocity between the gas and particles, thereby, tend to be destroyed due to the increasing gas flow within the cluster. Interactions between gas and particles promote energy dissipation component by drag forces due to the velocity difference between the gas phase and the solids phase.

When the energy dissipation component $\Omega_{dr}$ is the same order as the collisional energy dissipation component $\Omega_{co}$, the heterogeneous structure is compromised by the gas-particles interactions and the interactions of collisions of particles, and identifies a collision-hydrodynamic-dominant (CHD) heterogeneous structure. The heterogeneous structures are formed with the variation of drag energy dissipation component and collisional energy dissipation component in the computational cell.

In order to identify the contributions of the hydrodynamic interactions and the interactions of collisions on flow of heterogeneous structures, the dominant mechanism of heterogeneous structure is identified as follows:

$$\Phi_{\iota} = \frac{\Omega_{dr,\min}}{\Omega_{co,\min}} = \begin{cases} \text{HD} & \Omega_{dr,\min}/\Omega_{co,\min} \geq A_1 \\ \text{CHD} & A_1 < \Omega_{dr,\min}/\Omega_{co,\min} < A_2 \\ \text{CD} & \Omega_{dr,\min}/\Omega_{co,\min} \geq A_2 \end{cases} \tag{3.64}$$

where $A_1$ and $A_2$ are the base line values. Here, we assume that the base lines of $A_1$ and $A_2$ are 10.0 and 0.1, respectively. The variations of dominant mechanisms of the HD, CD and CHD regimes are shown in Fig. 3.11 in the riser. The controlling mechanism index $\Phi_i$ represents an occurrence of the controlling mechanism $i$ in the computational cells. The distributions of instantaneous controlling mechanism index are shown in Fig. 3.11 in the riser. The variation of instantaneous controlling mechanism index suggests that the formation of clusters is dominated by hydrodynamics interactions and interactions of collisions of particles in the riser.

The fraction of the controlling mechanism $\psi_i$ defines the ratio of the occurrence number of $\Phi_i$ to the total number $N$ of computational cells of the riser.

$$\psi_{\iota} = \frac{\sum \Phi_i}{N} \tag{3.65}$$

where the subscript $i$ represents HD, CHD, CD and HF, respectively. The sum of the fractions of $\psi_i$ is 1.0. The instantaneous fractions of the controlling mechanism of heterogeneous structures are shown in Fig. 3.12 in the riser. The controlling mechanism CD of heterogeneous structures is lowest, and the next in importance is the controlling mechanism HD. The fractions of the controlling mechanism CHD of heterogeneous structures is largest, indicating the heterogeneous structures arise as

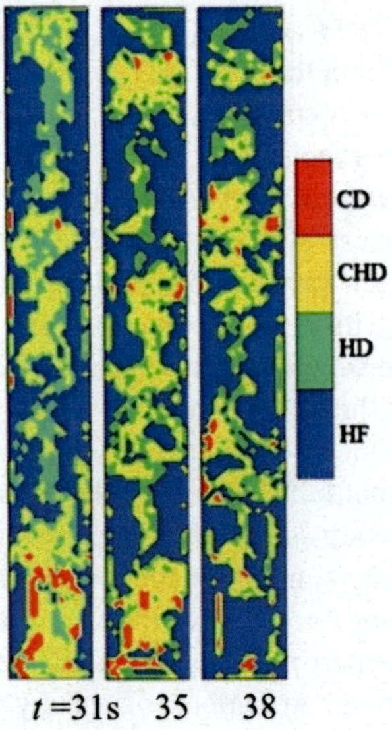


**Fig. 3.11** Instantaneous dominant mechanisms for existence of heterogeneous structures

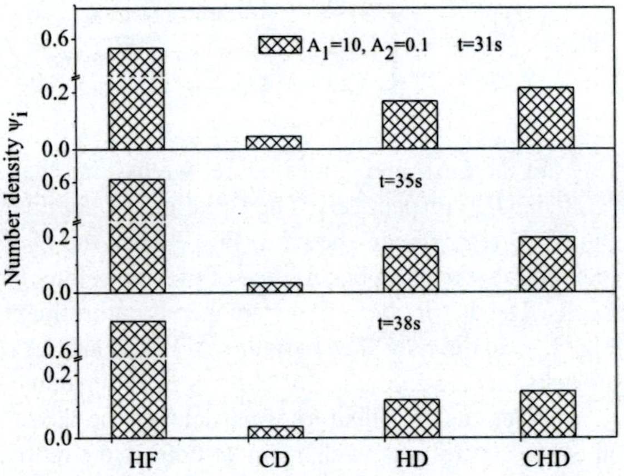


**Fig. 3.12** Profile of number density of cluster dominant mechanisms in the riser

a result of two controlling mechanisms operating in parallel for the inelastic collisions of particles and the hydrodynamic interactions. Forthermore, it indicates that the formation of the heterogeneous structures is contributed to the compromise of competition between the interactions of collisions of particles and the hydrodynamic interactions of gas-particles mixture.

### 3.6.9 Predicted Cluster Properties from DCSD Drag Model

The distributions of cluster solids volume fractions $\varepsilon_{s,c}$ are shown in Fig. 3.13 with the change of volume fractions of particles $\alpha_s$ of computational cells. Both simulations and calculations using empirical correlations proposed by Gu and Chen (1998) and Harris et al. (2002) show the cluster solids volume fractions increase with the increase of volume fractions of particles. The simulated cluster solids volume fractions are larger than calculations using empirical correlations because they were correlated from experimental data measured near the walls.

The distributions of cluster velocities are shown in Fig. 3.14 using DCSD drag model as a function of volume fraction of particles. The cluster velocities are negative at high volume fraction of particles. The measured cluster velocities at the riser wall have been reported, and the values fall in the range 0.5–2 m/s irrespective of the

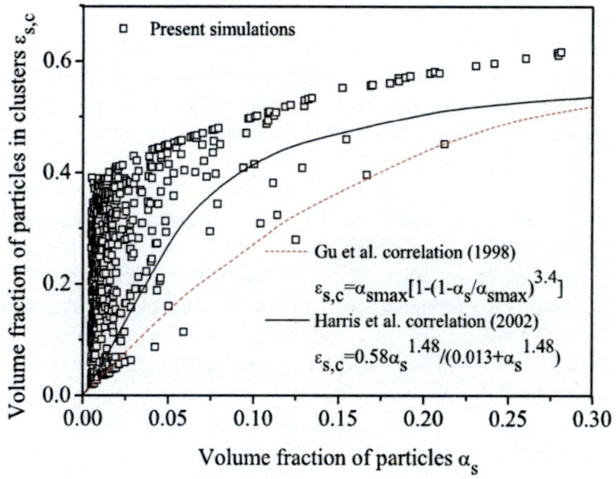


**Fig. 3.13** Comparison of solids volume fractions of clusters

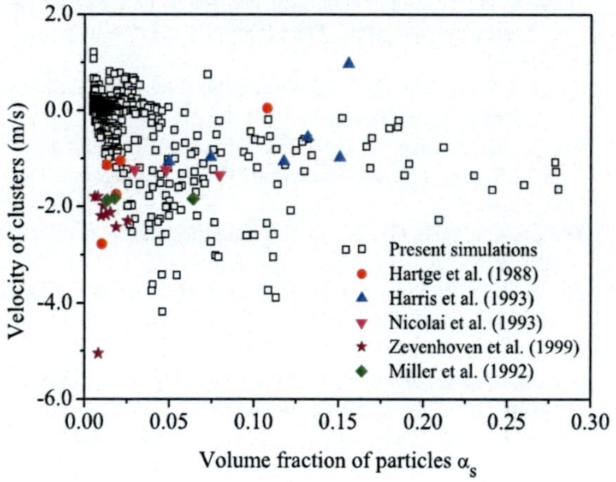


**Fig. 3.14** Comparison of cluster velocity using DCSD drag model and experiments

experimental measurement conditions (Miller and Gidaspow 1992; Hartge et al. 1988; Nicolai et al. 1993; Zevenhoven et al. 1999). The measured cluster velocity at the riser walls appeared insensitive to change in cross sectional averaged volume fraction of particles. Both simulations and experiments show the downward clusters with high solids volume fractionsare near the riser walls. On the other hand, the upward cluster velocities are positive at low volume fractions of particles in the center of the riser.

## References

Agrawal, K., Loezos, P. N., Sundaresan, S. M., & S, . (2001). The role of meso-scale structures in rapid gas-solid flows. *Journal of Fluid Mechanics, 445,* 151–185.

Dan, L., Shuyan, W., Guodong, L., Huilin, L., Xiaoxue, J., Ming, T., & Zhenjie, L. (2018). A dynamic cluster structure-dependent drag coefficient model applied to gas-solid risers. *Powder Technology, 325,* 381–395.

Ergun, S. (1952). Fluid flow through packed columns. *Chemical Engineering Progress, 48,* 89–94.

Estejab, B., & Battaglia, F. (2016). Assessment of drag models for Geldart A particles in bubbling fluidized beds. *Transactions of the ASME of Journal of Fluids Engineering ,138,* 031105 1–12

Fox, R. O. (2014). On multiphase turbulence models for collisional fluid-particle flows. *Journal of Fluid Mechanics, 742,* 368–424.

Garside, J., & Al-Dibouni, M. R. (1977). Velocity-voidage relationships for fluidization and sedimentation in solid-liquid systems. *Industrial & Engineering Chemistry Process Design and Development, 16,* 206–214.

Gibilaro, G. I., Felice, D. R., Waldram, P. S., & Foscolo, U. P. (1985). Generalized friction factor and drag coefficient correlations for fluid-particle interactions. *Chemical Engineering Science, 40,* 1817–1823.

Gidaspow, D. (1994). *Multiphase flow and fluidization: Continuum and kinetic theory descriptions.* New York: Academic Press.

Gidaspow, D., & Ettehadieh, B. (1983). Fluidization in two-dimensional beds with a jet, Part II: Hydrodynamics modeling. *Industrial and Engineering Chemistry Fundamentals, 22,* 193–201.

Gidaspow, D., & Huilin, L. (1998). Equation of state and radial distribution functionsof FCC particles in a CFB. *AIChE Journal, 44,* 279–293.

Good, G. H., Gerashchenko, S., & Warhaft, Z. (2012). Intermittency and inertial particle entrainment at a turbulent interface: The effect of the large-scale eddies. *Journal of Fluid Mechanics, 694,* 371–398.

Grace, J. R., Avidan, A. A., & Knowlton, T. M. (1997). *Circulating fluidized beds.* Springer: Blackie Academic and Professional.

Gu, W. K., & Chen, J. C. (1998). A model for solid concentration in circulating fluidizedbeds. In L. S. Fan & T. M. Knowlton (Eds.), *Fluidization* (Vol. IX, p. 501). Colorado: Engineering Foundation.

Haff, P. K. (1983). Grain flow as a fluid mechanical phenomenon. *Journal of Fluid Mechanics, 134,* 401–948.

Harris, A. T., Davidson, J. F., & Thorpe, R. B. (2002). The prediction of particle cluster properties in the near wall region of a vertical riser. *Powder Technology, 127,* 128–143.

Hartge, E. U., Resneer, D., & Werther, J. (1988). Solids concentration and velocitypatterns in circulating fluidized beds. In P. Basu & J. F. Large (Eds.), *Circulating Fluidized Bed Technology II* (pp. 165–180). Oxford: Pergamon.

Huilin, L. (2017). *Kinetic theory for dense fluid-solid flow.* Science Press

Huilin, L., & Gidaspow, D. (2003). Hydrodynamics of binary fluidization in a riser: CFD simulation using two granular temperatures. *Chemical Engineering Science, 58,* 3777–3792.

Huilin, L., Gidaspow, D., Bouillard, J., & Wentie, L. (2003). Hydrodynamic simulation of gas-solid flow in a riser using kinetic theory of granular flow. *Chemical Engineering Journal ,95,* 1–13.

Igci, Y., Andrews, A. T., Sundaresan, S., Pannala, S., & O'Brien, T. (2008). Filtered two-fluid models for fluidized gas-particle suspensions. *AIChE Journal, 54,* 1431–1448.

Igci, Y., & Sundaresan, S. (2011). Constitutive models for filtered two-fluid models of fluidized gas-particle flows. *Industrial and Engineering Chemistry Research, 50,* 13190–14132.

Kotz, S., & Nadarajah, S. (2000). *Extreme value distributions, theory and applications.* London: Imperial College Press.

Li, J., & Kwauk, M. (2003). Exploring complex systems in chemical engineering-the multi-scale methodology. *Chemical Engineering Science, 58,* 521–535.

Libby, P. A. (1975). On the prediction of intermittent turbulent flows. *Journal of Fluid Mechanics, 68,* 273–295.

McMillan, J., Shaffer, F., Gopalan, B., Chew, J. W., Hrenya, C. M., Hays, R., Karri, S. B. R., & Cocco, R. (2013). Particle cluster dynamics during fluidization. *Chemical Engineering Science, 100,* 39–51.

Miller, A., & Gidaspow, D. (1992). Dense vertical gas-solid flow in a pipe. *AIChE Journal, 38,* 1801–1815.

Nicolai, R., Goedicke, F., Tanner, H., & Reh, L. (1993). Particle induced heattransfer between walls and gas-solid fluidized beds. In A. Avidan (Ed.) *Circulating Fluidized Bed Technology IV*, AIChE New York, pp. 305–309.

Ozarkar, S. S., Yan, X., Wang, S., Milioli, C. C., Milioli, F. E., & Sundaresan, S. (2015). Validation of filtered two-fluid models for gas-particle flows against experimental data from bubbling fluidized bed. *Powder Technology, 284,* 159–169.

Richardson, J. F., & Zaki, W. N. (1954). Sedimentation and fluidization: Part I. *Transactions of the Institution of Chemical Engineers, 32,* 35–53.

Rowe, P. N. (1961). Drag forces in a hydraulic model of a fluidized bed. Part 2. *Transactions of the Institution of Chemical Engineers, 39,* 175–180.

Sarkar, A., Miloioli, F. E., Ozarkar, S., Li, T., Sun, X., & Sundaresan, S. (2016). Filtered sub-grid constitutive models for fluidized gas-particle flows constructed from 3-D simulations. *Chemical Engineering Science, 152,* 443–456.

Sharma, A. K., Tuzla, K., Matsen, J., & Chen, J. C. (2000). Parametric effects of particle size and gas velocity on cluster characteristics in fast fluidized beds. *Powder Technology, 111,* 114–122.

Shuai, W., Guangbo, Z., Guodong, L., Huilin, L., Feixiang, Z., & Tianyu, Z. (2014). Hydrodynamics of gas-solid risers using cluster structure-dependent drag model. *Powder Technology, 254,* 214–227.

Shuai, W., Huilin, L., Guodong, L., Zhiheng, S., Pengfei, X., & Gidaspow, D. (2011). Modeling of cluster structure-dependent drag with Eulerian approach for circulating fluidized beds. *Powder Technology, 208,* 98–110.

Soong, C. H., Tuzla, K., & Chen, J. C. (1995). Experimental determination of cluster size and velocity in circulating fluidized bed. In J. F. Large & C. Laguerie (Eds.), *Fluidization VIII* (pp. 219–227). New York: Engineering Foundation.

Syamlal, M., & O'Brien, T. J. (1987). *A generalized drag correlation for multiparticle systems.* Department of Energy, Office of Fossil Energy, Morgantown Energy TechnologyCenter: Report U.S.

Syamlal, M., Rogers, W., & O'Brien, T. J. (1993). MFIX documentation theory guide. DOE/METC-94/1004 NTIS/DE94000087 Department of Energy, Office of Fossil Energy, Morgantown Energy Technology Center Morgantown, WV, Springfield, VA.

Tuzla, K., Sharma, A. K., & Chen, J. C. (1998). Transient dynamics of solid concentration in downer fluidized bed. *Powder Technology, 100,* 166–172.

Wen, C. Y., & Yu, Y. H. (1966). Mechanics of Fluidization. *American Institute of Chemical Engineers Symposium Series, 62,* 100–111.

Wylie, J. J., & Koch, D. L. (2000). Particle clustering due to hydrodynamic interactions. *Physics of Fluids, 12,* 964–970.

Yin, X., Zenk, J. R., Mitran, P. P., & Hrenya, C. M. (2013). Impact of collisional versus viscous dissipation on flow instabilities in gas-solid systems. *Journal of Fluid Mechanics, 727,* 1–12.

Zevenhoven, R. (1999). Near wall particle velocity and concentration measurementsin circulating fluidized beds in relation to heat transfer. 15th International Conference on Fluidized Bed Combustion, Savannah GA (ed) American Society of Mechanical Engineers, Paper No. FBC99–0097.

# Chapter 4
# Experimental Foundation

## Nomenclature

| | |
|---|---|
| $d_p$ | Particle diameter, m |
| $e$ | Coefficient of restitution |
| $f$ | Frequency, Hz |
| $g_0$ | Radial distribution function at contact |
| $g$ | Gravity, $m/s^2$ |
| $G_s$ | Solids mass flux, $kg/m^2{\cdot}s$ |
| $I$ | Intensity of radiation, counts/s |
| $k$ | Mass attenuation coefficient, $m^2/kg$ |
| $l$ | Path length, m |
| $L$ | Distance, m |
| $N$ | Number of particles |
| $p$ | Gas pressure, Pa |
| $p_s$ | Solids pressure, Pa |
| $R$ | Riser radius, m |
| $v$ | Velocity, m/s |
| $W$ | Potential, $m^2/s^2$ |

## Greek Symbols

| | |
|---|---|
| $\alpha$ | Volume fraction |
| $\rho$ | Density of material, $kg/m^3$ |
| $\sigma$ | Standard deviation of velocity, $m^2/s^2$ |
| $\theta$ | Granular temperature, $m^2/s^2$ |
| $\mu$ | Viscosity, Pa·s |
| $\eta$ | Alumina extraction efficiency |



H. Lu et al., *Computational Fluid Dynamics and the Theory of Fluidization*,
https://doi.org/10.1007/978-981-16-1558-0_4

**Subscripts**

*g* Gas phase
*s* Solids phase

## 4.1 Introduction

Fluidized bed reactors feature a dynamic mixture of gas and solids, causing that the proper measurement and monitoring of hydrodynamic phenomena is a challenging endeavor. Several intrusive and non-intrusive measurement techniques are used, and they look into distinct physical properties of the fluidized bed related to the presence of bubbles and clusters.

The radiography is conducted with X-ray and gamma-ray in fluidized beds. The radio source is positioned on one side of the bed, while a receiver is placed at the opposite side. The main advantage of radiography is its non-intrusive character and the good visual interpretation. These advantages are, however, outnumbered by many disadvantages. For example, the detection range is limited by the power of the radiation source (~30 cm), thus making radiography not applicable in large units. Moreover, stringent safety regulations make radiography expensive.

The optical fiber probe consists of a glass fiber, sends light into the dense phase of a fluidized bed. Depending on the configuration, either the reflection or the transmission of the light in the dense phase is measured. For reflection probes, the sending and receiving optics are mounted at the same side of the measurement volume, while for the transmissive configuration, the sending and receiving end face each other, with the measurement volume in between. On the other hand, the small changes in the gas volume fractions of a fluidized bed can be seen as changes to the capacitance in the bed. Hence, measuring the capacitance is a simple and accurate means to describe the hydrodynamics in fluidized beds. The capacitance probing relies on two closely spaced electrodes reaching into the bed and measuring in a small and defined volume. Compared to other measurement techniques, the optical probes and capacitance probing have the advantage of being able to inexpensively and very locally measure bed properties. Due to their intrusive nature, the bed hydrodynamics may be disturbed. However, with an appropriate design and positioning of the probes, these intrusive effects can be minimized. Generally, the optical probes and capacitance probing are primarily used in cold and non-reactive systems. Thus, they are mostly applied along with other measurement techniques (e.g., pressure fluctuation measurements) as a valuable comparison or calibration aid.

The tomography is a non-intrusive technique, which can be applied for the reconstruction of the cross-sectional distribution of solids volume fractions in a fluidized bed. Principally, the tomography measurements are mostly applied by means of high-energy photons emitted from X-ray or $\gamma$-ray sources, traveling through the fluidized bed. On the other hand, the electric capacitance tomography (ECT) uses a number of plate electrodes which are placed at a given height around the fluidized bed. For the measurement protocol, the electrodes are excited one after the other and the capacitance between the excited electrode and the remaining ones is measured. In this way, the electrical field is rotated around the column cross section in discrete steps. The ECT features the advantage of being fast (i.e., up to 1,000 frames/s), yet the image reconstruction is cumbersome, and the spatial resolution is limited.

The tracking individual particles, such as computer automated radioactive particle tracking, positron emission particle tracking, particle image velocimetry, or laser doppler velocimetry. Quantities that need to be measured in gas fluidized beds include solids volume concentrations, solids velocities and solids mass flows, the vertical and horizontal distribution of solids inside the bed, the lateral distribution of the fluidizing gas, temperatures and gas concentrations.

## 4.2 Solids Volume Fractions Measurements

### *4.2.1 Radial Solids Volume Fractions and Density*

Figure 4.1 shows a circulating fluidized bed (CFB) unit (Miller and Gidaspow 1992). Air and particles enter the CFB through the bottom U-tube whose diameter is smaller than that of the riser. Its inner diameter is 5.2 cm. The particle feed rate is controlled by a slide valve. The riser is 7.5 cm in ID and 6.58 m long, and particle size distribution has an average of 75 μm in diameter. The outflow stream enters a primary cyclone where the gas-particle separation process mainly occurs. Additional separation is performed in the secondary cyclone. Particles are recycled to a storage PVC hopper.

The solids volume fractions were obtained with an X-ray densitometer, seeing Fig. 4.2. The X-ray densitometer was used to measure the particle concentration (Seo and Gidaspow 1987; Miller and Gidaspow 1992). The source of the X-ray densitometer was a 200-mCi Cu-244 source having 17.8 half-life. It emitted X-rays with a photon energy between 12 and 23 keV. The source was contained in a ceramic enamel, recessed into a stainless-steel support with a tungsten alloy packing, and sealed in a welded capsule. The intensity of the transmitted X-ray beam was detected by the NaI crystal scintillation detector. It consisted of a 2-mm thick, 5.08-cm-diameter tube with a 0.13-mm-thick beryllium window. The photomultiplier of the detector was connected sequentially to a preamplifier, an amplifier, and a double channel analyzer. The amount of radiation that is absorbed by a material can be given by the Beer-Bougert-Lambert law.

$$I = I_0 \exp(-k\rho l) \tag{4.1}$$

where $I$ is the intensity of transmitted radiation, $I_o$ is the intensity of incident radiation, $k$ is the attenuation coefficient, $\rho$ is the density of material, and $l$ is the path length. The logarithmic form of the equation for gas–solid phase fluidized beds is

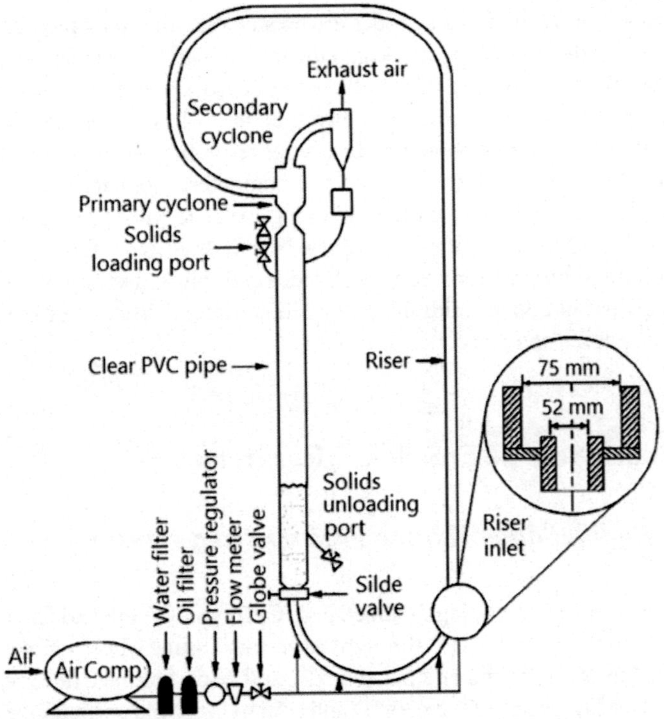


**Fig. 4.1** Circulating fluidized bed (CFB)

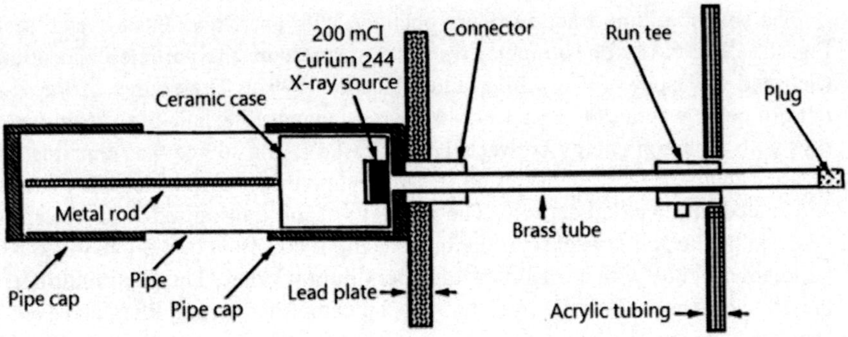


**Fig. 4.2** X-ray probe to measure particle concentration

$$\ln(I/I_0) = [k_g \rho_g (1 - \alpha_s) + k_s \rho_s \alpha_s] l \tag{4.2}$$

The attenuation coefficient for the 75 μm FCC particles was found to be 0.08133 $cm^2/g$. The solids volume fraction $\alpha_s$ is determined according to measured intensity of transmitted radiation.

The instantaneous gas volume fractions are shown in Fig. 4.3 at superficial gas velocity $v_g$ and solids mass fluxes $G_s$ of 2.67 m/s and 28.17 kg/m$^2$s, respectively. The data acquisition sampling frequency was set to 40 Hz. To eliminate any spurious effects of initial conditions on the data analysis, data taken during the first second were systematically eliminated from time series data sets (Huilin et al. 1997). The large oscillation of gas volume fractions is found because of the formation of clusters in the dense and lower section of the riser.

Time-average data were obtained for flow of 75 μm FCC catalyst particles in a 7.5 cm diameter riser as a function of gas and solids flow rates. Particle concentrations are shown in Fig. 4.4. We see a dilute core and the central dense annulus.

Figure 4.5 shows the distribution of cross-sectional solids density at the height of 4.18 m of the riser. The spatial resolution and image reconstruction of solids phase density is achieved with X-ray tomography. From measured local porosity,

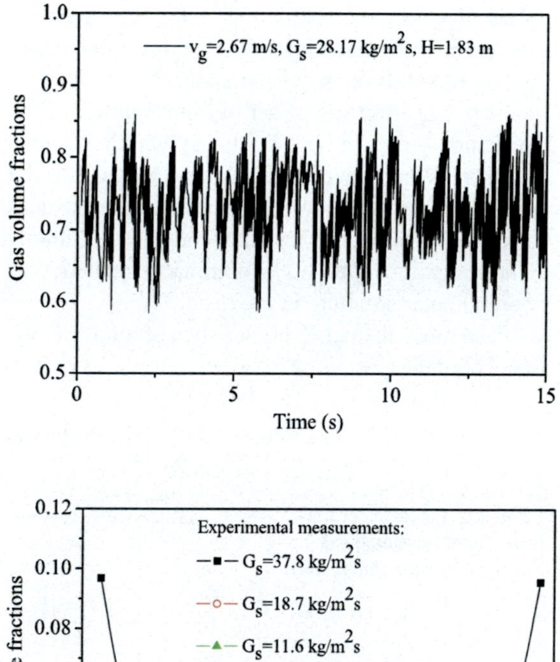


**Fig. 4.3** Instantaneous gas volume fraction fluctuations by X-ray densitometer

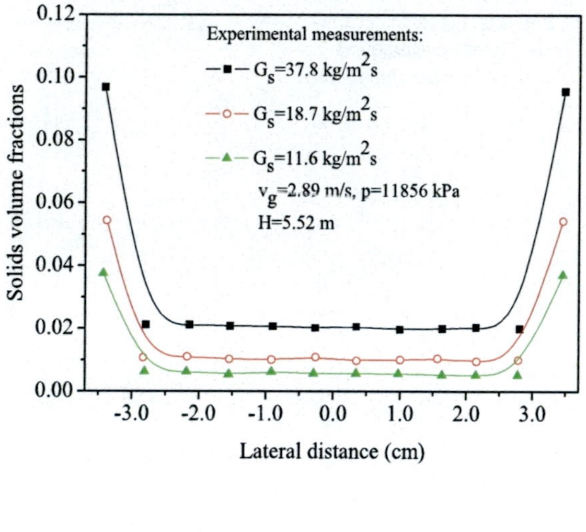


**Fig. 4.4** Particle concentrations with a dilue core and dense layer at the wall

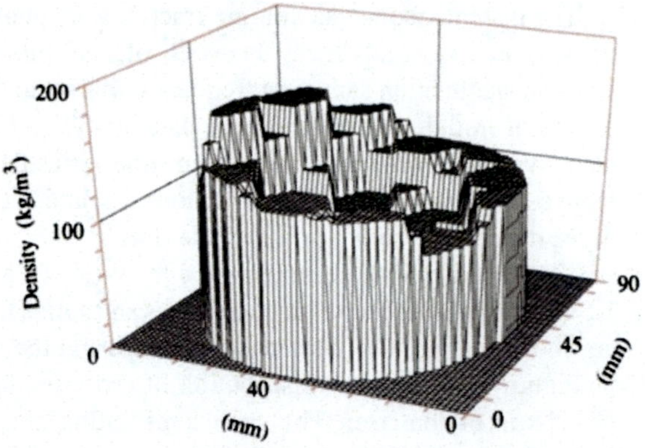


**Fig. 4.5** Distribution of solids density at the height of 4.18 m

the cross-sectional distribution of solids density is constructed by means of the iterative algebraic reconstruction techniques. Measured solids densities show clearly the asymmetrical behavior. This is due to the elbow type outlet. The cross-sectional averaged solids density is 86.2 kg/m$^3$.

A power spectrum of density fluctuations can distinguish between a well defined structure, such as a large bubble in bubbling fluidized beds or a core-annular regime in risers. When a well defined structure exits, a sharp peak is obtained for the spectrum. Figure 4.6 shows the power spectrum densities corresponding to the fluctuations of measured instantaneous gas volume fractions (Gidaspow et al. 2001). The diagram highlights a dominant frequency at about 0.18 Hz identifying the core-annular structure in the riser.

The most distinct characteristics of fluidized beds are the large-scale oscillations: bubbles and clusters. Figure 4.7 shows a comparison of the dominant frequency measured in the riser for flow of 75 μm particles using an X-ray densitometer to an analytical solution obtained from the basic equations of motion, making many

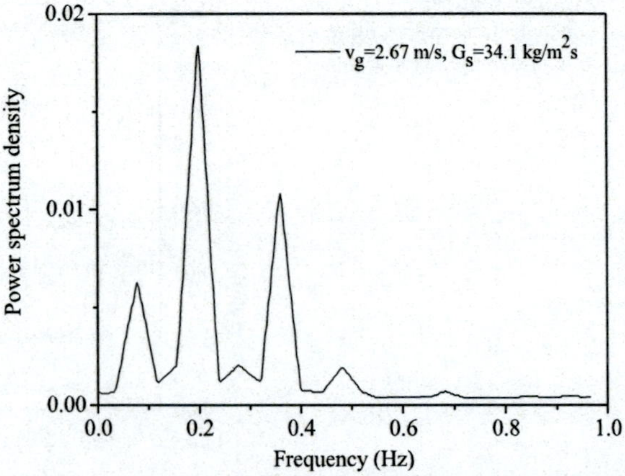


**Fig. 4.6** Power spectrum density of measured gas volume fraction fluctuations

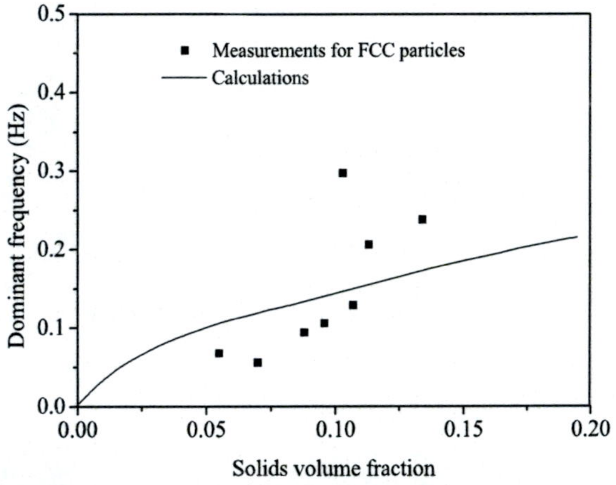


**Fig. 4.7** A comparison of riser dominant frequency to analytical solution

approximations. The measured dominant frequency increases with increasing solids volume fractions. The frequencies oscillation was obtained from momentum balance of particles in the vertical pipe. The analytical solution for the frequency is (Gidaspow and Mostofi 2003)

$$f = \frac{1}{2\pi}\left(\frac{g}{H_0}\right)^{1/2}\left[\frac{(3\alpha_s/\alpha_g + 2)\alpha_s}{\alpha_{s0}}\right]^{1/2} \tag{4.3}$$

where $\alpha_{s0}$ and $H_0$ are some initial bed volume fraction and height of bed filled with particles. The basic frequency is the gravity divided by initial bed height. It is corrected for by approximately the square root of the solids volume fraction. This gives a lower fundamental frequency in the riser.

### *4.2.2 Solids Volume Fractions of Bubbling Fluidized Bed*

The electric capacitance tomography (ECT) is physically based on the interactions of the gas-solids distribution with the electric field lines. The difference in the dielectric constant between solids and gas is measured by means of a number of plate electrodes which are placed at a given height around the column of the bubbling fluidized bed, seeing Fig. 4.8 (Shuai 2014). The diameter and height of the column are 60 mm and 1.2 m. The diameter and density of particles are 530 μm and 2650 kg/m$^3$. The static bed height of particles is 0.2 m. The electrodes are excited one after the other and the capacitance between the excited electrode and the remaining ones is measured.

The image reconstructions of solids volume fractions are shown in Fig. 4.9 at two different fluidizing gas velocities. Both cases show that the solids volume fractions

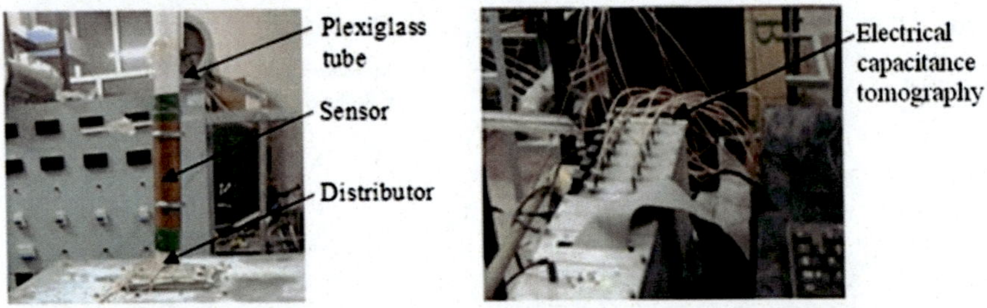


**Fig. 4.8** Experimental system of bubbling fluidized bed mounted with ECT sensor

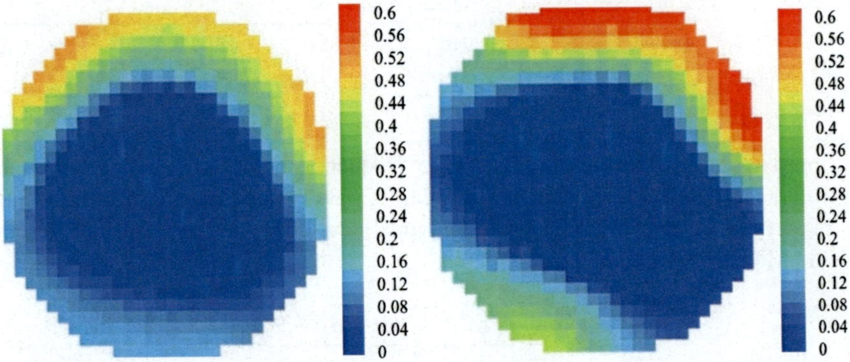


**Fig. 4.9** Reconstructed solids volume fractions (left: $v_g = 0.55$ m/s, right: $v_g = 0.75$ m/s) (z = 0.2 m)

are low at the center and high near the walls. Both measured particle flow patterns show a upward flow in the center region and downward flow near the wall, resulting in a good mixing of particles in the bed.

## 4.3 Measured Particles Velocities and Granular Temperature

### *4.3.1 Velocity Measurements of Particles*

Figure 4.10 illustrates the particle image velocity (PIV) method of obtaining instantaneous velocities of particles (Gidaspow and Huilin 1998). The high-resolution microimaging/measuring system for the particle-velocity measurements consist essentially of two units: a high-resolution microimage system and a data managing system. The high-resolution microimage system is a 2/3-in. (17-mm) color video camera (DXC-151A) that uses a charge coupled device (CCD), a solid-stage image

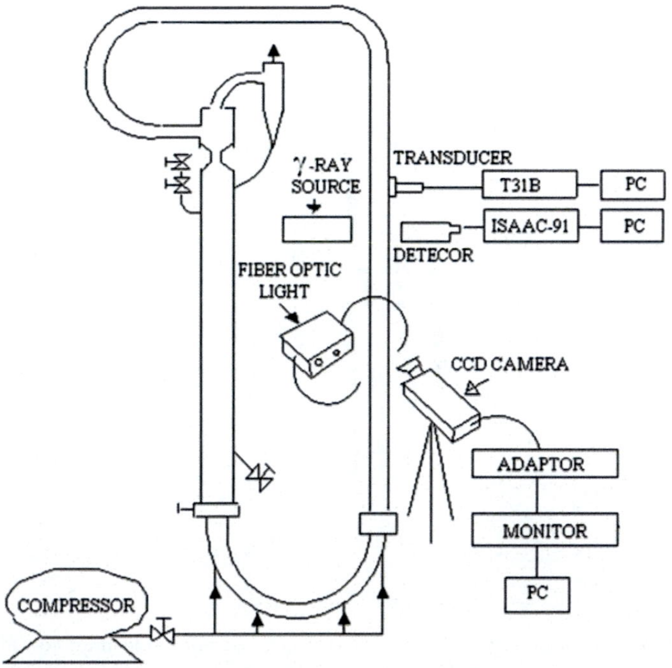


**Fig. 4.10** Circulating fluidized bed with particle pressure transducer, CCD camera and X-ray densitometer

sensor. This camera has ten electronic shutter settings and four modes for gain control. The horizontal resolution of the camera is 460 TV lines, and it has a sensitivity of 2000 lx at 0 dB for gain. The camera adapter is a Sony CMA-D2, which connects the camera to a 486/33-MHz IBM computer. The personal computer has a micro-imaging board and a micro-imaging software, Image-Pro Plus, for data measurement and analysis.

For good visualization of the microscopic movement of particles, a fiber-optic light was used. The area of view in most experiments was a 5 × 20 mm. As the particles were fluidized inside the riser, the camera with a zoom lens, 18–108 mm, and close-up focus transferred its field of view into the monitor. Figure 4.11 shows a typical streak made by the particles. These streak lines represent the distance traveled by the particles in a given time interval specified on the camera. The images were then captured and digitized by the Image-Pro Plus software. To obtain reliable velocity information, the time between exposures must be selected so as to obtain a sufficient displacement to achieve an acceptable velocity resolution, but must not be so large that the particle moves out of the field of view. The local velocity is then estimated from the equation

$$v = \frac{\Delta L(x, y)}{\Delta t} \tag{4.4}$$

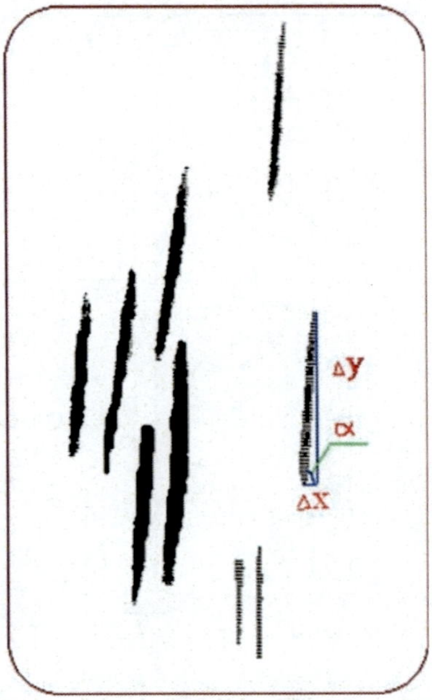


**Fig. 4.11** Streak images captured by the CCD camera recording system for flow of 75 μm FCC particles. This figure shows a small portion from screen on which images are captured, for example, $\Delta\, y = 1.44$ mm; $\Delta x = 0.22$ mm; $(\alpha = 80°; \Delta t = 0.5$ ms. Velocity is negative when streak is formed moving down on the computer monitor

Typical particle axial and radial velocity distributions are shown in Fig. 4.12. The ordinate, expressed in $(\text{cm/s})^{-1}$, is the probability that the velocity will fall within the indicated velocity range. It is obtained by dividing the number of particles in the velocity range shown by the total number of measurements and by the velocity in the indicated velocity range. Velocities are negative when the particles flow down, as visually observed.

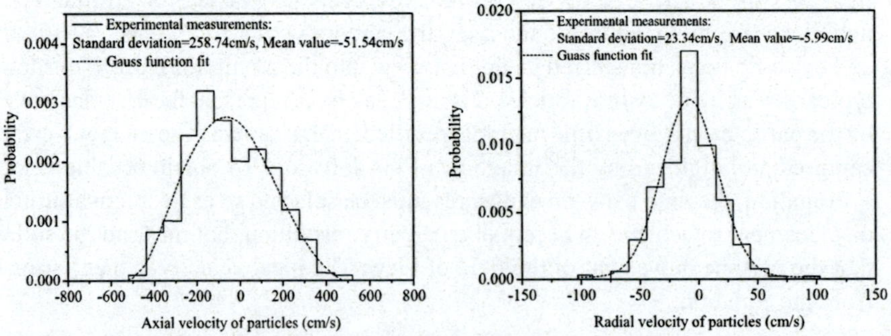


**Fig. 4.12** Histograms of axial and radial particle velocities

### 4.3.2 Granular Temperature of Particles

From measured axial and radial velocity of particles, the variances of velocity were calculated using the standard equation of statistics (Gidaspow and Huilin 1996).

$$\sigma_k^2 = \frac{1}{N}\sum_{i=1}^{N}(v_i - v_m)^2 \tag{4.5}$$

where $v_i$ is the velocity of particle $i$ in the $k$ direction and $v_m$ is the mean. Figure 4.13 shows the distribution of variances of axial and radial velocities component. Both axial variances and radial variances increase with increasing solids mass fluxes. The velocity components in the $x$ radial direction and $\theta$ direction are small compared to the $z$ direction, the vertical direction. Hence the variance in the $\theta$ direction is about one-tenth that in the principal direction of the flow. It is therefore reasonable to assume that the $x$ and $\theta$ variances are equal. Hence we compute the granular temperature as

$$\theta = \frac{1}{3}(\sigma_\theta^2 + \sigma_x^2 + \sigma_z^2) = \frac{2}{3}\sigma_x^2 + \frac{1}{3}\sigma_z^2 \tag{4.6}$$

The anisotropy of the velocity fluctuations is a strong function of the position, since granular temperature and the mean velocity are strongly correlated, as observed experimentally.

Figure 4.14 shows the distribution of granular temperature along radial direction of the riser. The profiles are not symmetrical. The granular temperature is high at the center and low near the walls of the riser.

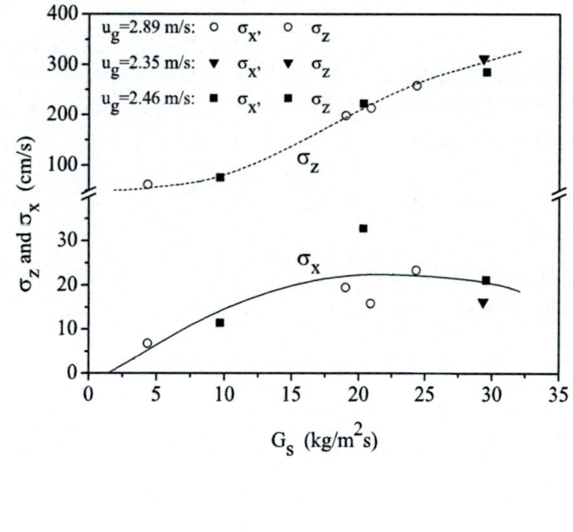


**Fig. 4.13** Distributions of variances of particles velocities

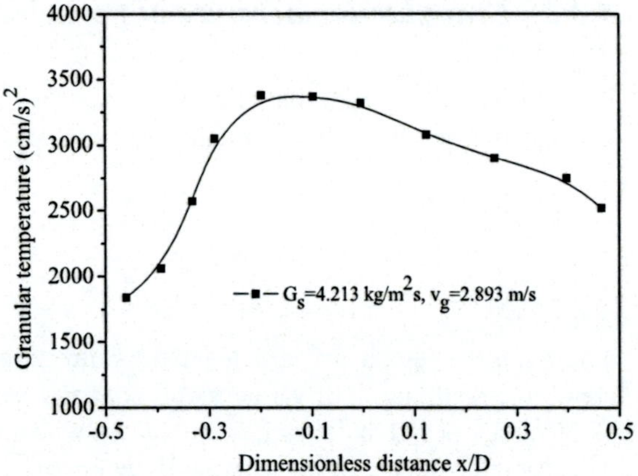


**Fig. 4.14** Granular temperatures along radial distance in the dilute region

The solids volume fractions were measured by gamma ray method. The gamma ray source is a 500-Mci-Cs-137 source having a single gamma ray of 667 keV and a half-life of 30 years. The Cs-137 source was sealed in a welded stainless steel capsule. The source holder was welded, filled with lead, and provided with a shutter to turn off the source (Gidaspow et al. 1983; Jung et al. 2005). The transmitted beam intensity was measured by using a NaI(Ti) crystal scintillation detector (EG & G Model 905), which was connected to photo multiplier base (EG & G Model 266). The signals from the photo multiplier were passed through a preamplifier, an amplifier, and a single-channel analyzer, a rate meter, and a data acquisition system. The rate meter has a selector and a 0–100-mV scale range. The beam from the source was collimated through a 1.27 cm hole on 3.8 cm thick lead plate.

Figure 4.15 shows distribution of granular temperature as a function of solids volume fractions. The granular temperature rises rapidly with solid-volume fraction, reaches a maximum near a solid-volume fraction of about 6%, and then decreases

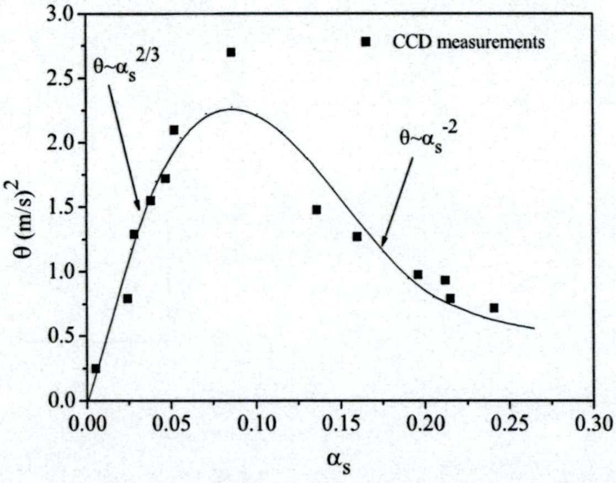


**Fig. 4.15** Granular temperatures for 75 μm FCC

slowly (Gidaspow et al. 2004). Such behavior is reasonable since the granular temperature is zero with no particles and will approach zero for a packed bed, in which there is no particle motion.

In the dilute flow without particle collisions, the mass, momentum, and energy equations for the particulate phase, as a first approximation, using the Maxwellian velocity distribution, are (Gidaspow 1994; Huilin 2017):

$$\frac{D\alpha_s}{Dt} = -\alpha_s \frac{\partial \mathbf{v}_s}{\partial \mathbf{r}} \quad \text{(constant } \rho_s\text{, conservation of particles)} \tag{4.7}$$

$$\frac{D\mathbf{v}_s}{Dt} = -\frac{1}{\alpha_s \rho_s} \frac{\partial \mathbf{p}}{\partial \mathbf{r}} \quad \text{(inviscid particle momentum balance)} \tag{4.8}$$

$$\frac{3}{2}\alpha_s \rho_s \frac{D\theta}{Dt} = -\alpha_s \rho_s \frac{\partial \mathbf{v}_s}{\partial \mathbf{r}} \quad \text{(kinetic energy balance with no dissipation)} \tag{4.9}$$

Combination of the energy conservation equation with the conservation of the mass equation, gives

$$\frac{3}{2\theta}\frac{D\theta}{Dt} = \frac{1}{\alpha_s}\frac{D\alpha_s}{Dt} \tag{4.10}$$

This equation shows that the granular temperature is proportional to the two-thirds power of the solid-volume fraction, as shown below:

$$\theta \propto \alpha_s^{2/3} \tag{4.11}$$

Equation (4.11) is analogous to the isentropic temperature variation with density for an ideal gas. Such an assumption is a good first approximation in design of compressors and particularly expansion devices. It suggests that the granular temperature rises with the solids volume fraction due to the compression effect. Hence, the viscosity in the dilute limit will rise as $\alpha_s^{1/3}$ . It will not give the infinite value at $\alpha_s = 0$ obtained in the incompressible shear flow solution. There is no singularity.

For dense suspensions with no dissipation in the fluid and by the wall, the production-dissipation balance gives an equation for granular temperature as (Savage 1983; Gidaspow 1994)

$$(1-e^2)^{1/2}\theta^{1/2} = \left(\frac{5\pi}{18}\right)\ell\left(\frac{\partial v_s}{\partial y}\right) \tag{4.12}$$

As the mean free path, *l*, decreases with a higher volume fraction, the granular temperature drops. Equations (4.11) and (4.12) explain the behavior of Fig. 4.15.

## 4.4 Radial Distribution Function Measurement

Radial distribution functions play a key role in the statistical mechanics of liquids. They are often measured using neutron scattering techniques and light scattering techniques. We used our CCD camera system to measure the local particle distributions. Figure 4.16 shows a typical particle distribution image for 75 μm FCC particles, approximately at the same location as the particle volume fraction, granular temperature, and particle pressure measurements. The superficial gas velocity and solid mass flux are 2.46 m/s and 18.44 kg/m$^2$s, respectively. The CCD camera system stores the coordinates of each particle in the image. Its center position and area were determined by software IPPLUS and stored in a computer (Gidaspow and Huilin 1998).

The radial distribution function $g(r)$ was calculated as follows:

$$\Delta N = 2\pi r \frac{N}{AREA} g(r)\Delta r \tag{4.13}$$

where $N$ is the total particle number in the AREA, and $\Delta N$ is the particle number in the computing area. Thus, the radial distribution function is

$$g(r) = \frac{\Delta N}{N}\frac{AREA}{2\pi r \Delta r} \tag{4.14}$$

and the local density of particle is

$$\text{Local density of particle} = \left(\frac{N}{AREA}\right) g(r) \tag{4.15}$$

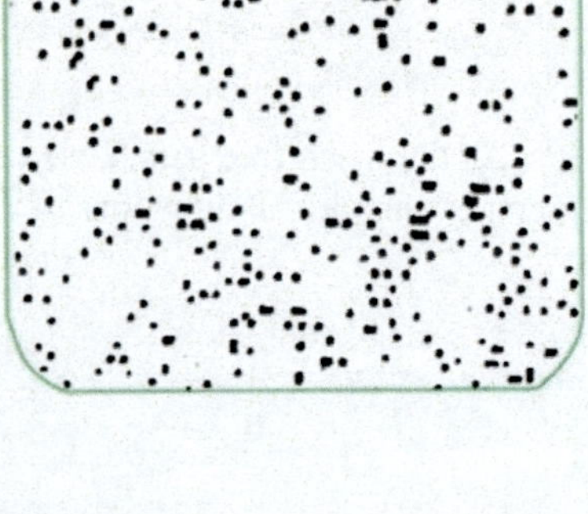

Fig. 4.16 A small portion of an image of particle distribution captured by CCD camera

Fig. 4.17 Calculation of radial distribution function

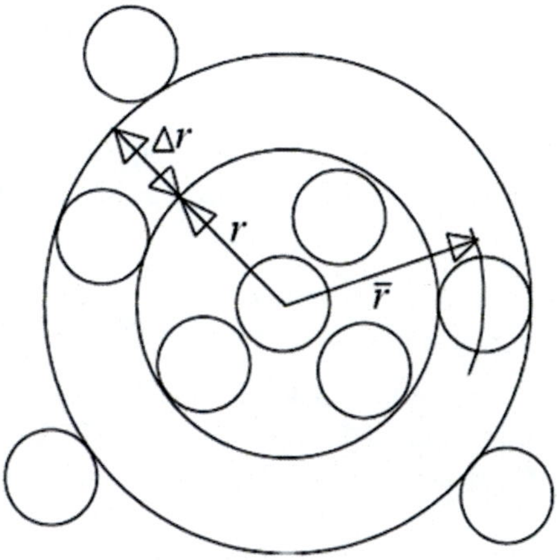


$$r = \sqrt{\left(x_i - x_j\right)^2 + \left(y_i - y_j\right)^2} \tag{4.16}$$

This method is very similar to the Monte Carlo simulation of the equation of state. The termination effect was eliminated by the boundary-condition treatment which is different from the periodicity assumption in molecular simulations. Figure 4.17 summarizes the concept of the radial distribution function and its calculation. To prevent two particles from being at the same location, for $r < d_p$ (particle size), $g(r) = 0$. As $r \rightarrow \propto$, local density = local density $\times g(r)$. Hence, $g(r) \rightarrow 1.0$.

Figure 4.18 shows the typical radial distribution function profiles for flow of 75 μm FCC particles. Compared with the hard-sphere model, we have nonzero values below the mean particle diameter because of the particle-size distribution. From these figures, the radial distribution function $g_0$ at contact, $r = d_p$, was determined.

A simple form for the radial distribution function, $g_0$, is that used by Bagnold (1954), as reviewed by Gidaspow (1994):

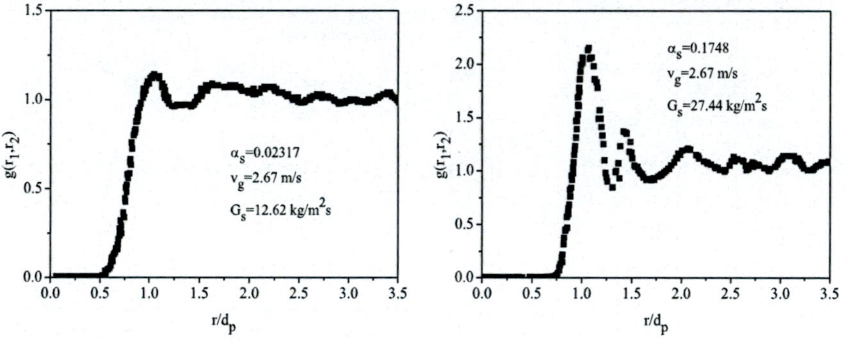


**Fig. 4.18** Profiles of radial distribution functions

$$g_0 = \left[1 - \left(\frac{\alpha_s}{\alpha_{\max}}\right)^{1/3}\right]^{-1} \quad (4.17)$$

Carnahan and Starling's equation is used to compare the experimental radial distribution function (Carnahan and Starling 1969), as follows:

$$g_0 = \frac{1}{1-\alpha_s} + \frac{3\alpha_s}{2(1-\alpha_s)^2} + \frac{\alpha_s^2}{2(1-\alpha_s)^3} \quad (4.18)$$

Figure 4.19 shows the comparison of the experimental data with computed results using Bagnold's equation (Bagnold 1954) and Carnahan and Starling's equation (Carnahan and Starling 1969). We see that the Bagnold equation agrees approximately with the data. A possible cause for the measured data being larger than that of Carnahan and Starling results is that the particle-size distribution is not uniform. The Carnahan and Starling equation is only suitable for monodisperse hard spheres. Qualitatively, the radial distribution function increases with increase in concentration of particles. We see that the Carnahan and Starling equation is not applicable to the dense flow of particles as the quantity of the radial distribution function at contact does not become infinite as particle concentration approaches the maximum packing value. The value of the radial distribution function can be calculated from the equation of particle collision frequency that was derived from the kinetic theory of granular flow (Gidaspow 1994).

Accurate characterization of radial distribution function is important because constitutive equations of solid phase derived from kinetic theory of granular flow depend on the radial distribution function. The radial distribution functions are computed using the hard-sphere discrete model in gas-particles fluidized bed (Huilin et al. 2005). The typical radial distribution function profiles in the fluidized bed are shown in Fig. 4.20 (Shuyan et al. 2009). The highest quantity of radial distribution

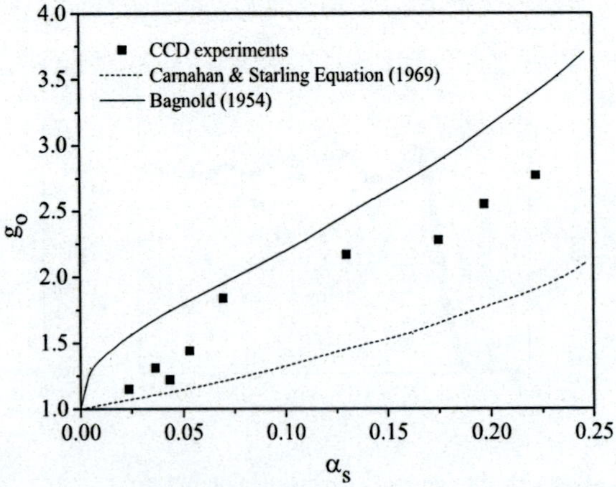


**Fig. 4.19** Measured and computed radial distribution function at particle contact

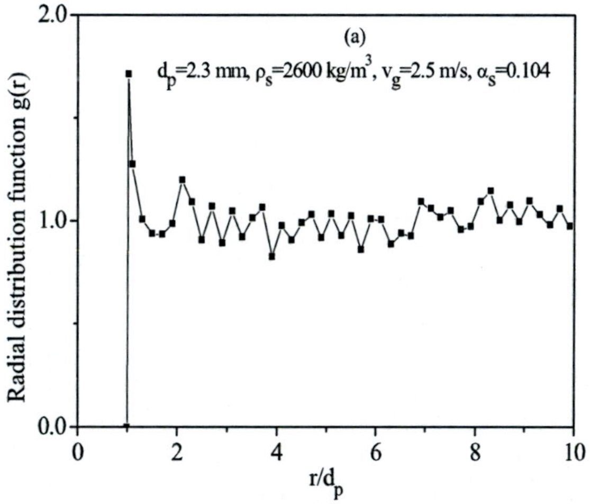


**Fig. 4.20** Calculated radial distribution function in fluidized bed

function is at the distance $r$ of a particle diameter, then decrease to unity as the $r/d_p$ number increases. The radial distribution function is a local measure of how close the particle distribution is to uniform (a distribution where all particle locations are equally likely). As $g(r)$ is greater or less than unity, it implies an enhanced or reduced probability relative to the uniform distribution. From these figures, the radial distribution function at contact ($r = d_p$), $g_o$, was determined.

Figure 4.21 shows the distributions of radial distribution function at contact as a function of concentration of particles. The radial distribution function increases with the increase of particle concentrations. We see that the calculated radial distribution function by Carnahan and Starling equation is lower than that by present simulations. For $\alpha_s > 0.55$, the Bagnold equation is applicable to dense flow as the quantity of

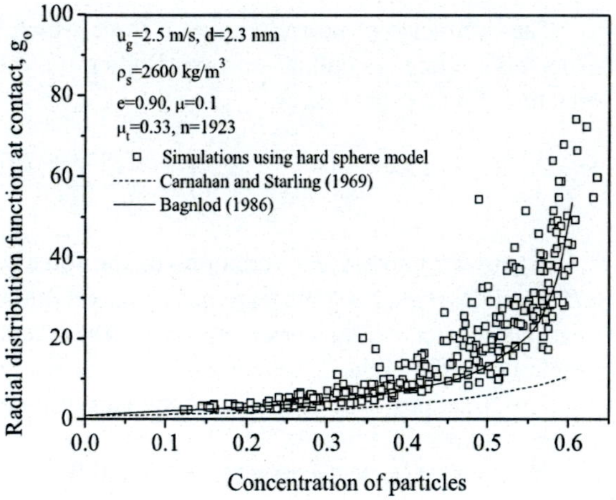


**Fig. 4.21** Calculated radial distribution functions using hard-sphere discrete model

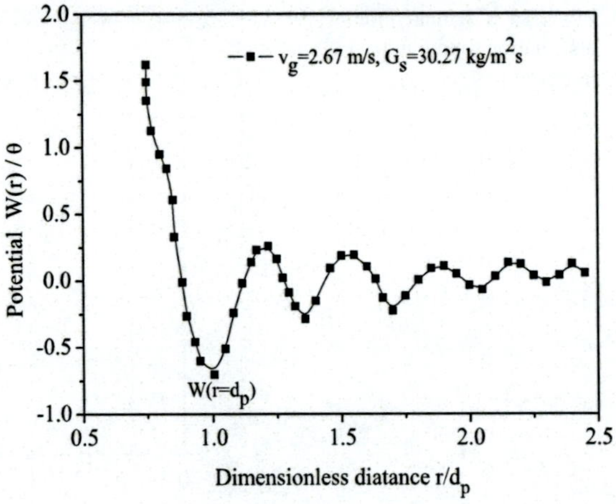


**Fig. 4.22** Variations of potential of mean force with dimensionless distance

the radial distribution function does go to infinite as particle concentrations approach to the maximum packing value.

The radial distribution function is the highest when the distance $r$ equals the particle diameter, decreases in an oscillating manner, and then goes to unity. The potential, $W$, was calculated from the Boltzmann relation. For the Boltzmann constant times the thermal temperature we have substituted the granular temperature which is the oscillating energy per unit mass of particles:

$$W(r) = -\theta \log[g(r)] \tag{4.19}$$

Figure 4.22 shows typical profiles of this potential for FCC particles divided by the granular temperature as a function of dimensionless distance. In these figures, $W(r)$ is named the structural energy barrier in colloid science. The potential is negative at the contact $r = d_p$.

The difference of potential gives a force which is called "structural force" and "depletion force" in colloid science (Hunter 1989; Balucani and Zoppi 1994). The relation can be expressed as:

$$F(r) = -\frac{d[W(r)/\theta]}{dr} \tag{4.20}$$

Figure 4.23 shows the variations of the force with dimensionless distance for 75 μm FCC particles. This force has its maximum value at $r_o$ and a minimum at the particle contact, $r = d_p$, respectively. At $r_o$ the potential is positive for FCC particles. Figure 4.24 shows the distribution of potential $W(r = d_p, \alpha_s)$ as a function of solids volume fractions. The potential at the particles contact decrease with increasing solids volume fractions.

If we approximate the measured potential $W(r = d_p, \alpha_s)$ by a Dirac delta function at contact, then the first peak of the radial distribution function gives the second term

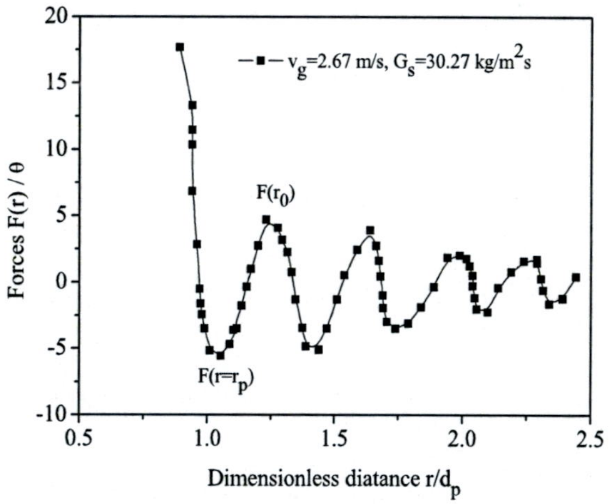


**Fig. 4.23** Profile of forces of FCC particles

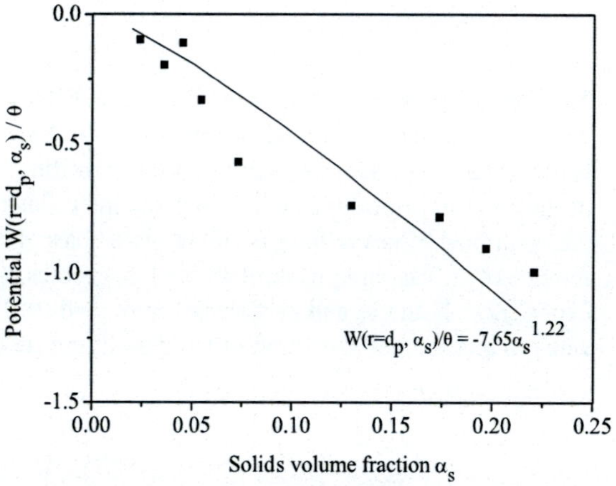


**Fig. 4.24** Variation of potential with solids volume fractions

of the equation of state, the collisional portion of particles (Egelstaff 1967; Hansen and McDonald 1991), as follows:

$$\frac{p_c}{\rho_s \alpha_s \theta} = -\alpha_s \frac{d[W(r = d_p, \alpha_s)/\theta]}{d\alpha_s} = 9.33\alpha_s^{1.22} \tag{4.21}$$

Figure 4.25 shows the comparison between predictions from Eq. (4.19) and the collisional portion of solids pressure $(1 + e)\ \alpha_s g_0$ with the restitution coefficient of 0.999 from kinetic theory of granular flow. We see that the collisional pressure calculated from the derivative of potential, $W$, that obtained from the first peak and the collisioanl portion of solids pressure from kinetic theory of granular flow are the same.

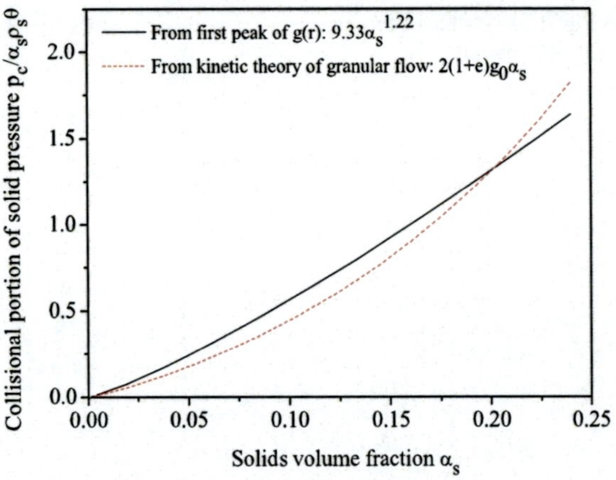


**Fig. 4.25** A comparison between the experimental data and predicted collisional portion

## 4.5 Particle-Pressure Transducer and Measured Particle Pressure

The particle pressure was measured using a differential pressure transducer PDCR made by Druck Incorporated, seeing Fig. 4.26. The operating principle of the pressure transducer is as follows (Gidaspow and Huilin 1998). The front of the diaphragm of the pressure transducer, which was mounted flush with the bed wall, experienced the combined forces of the gas and particle phase. Small passages about the circumference of the diaphragm admit air but no particles. Thus, the front of the diaphragm experiences both gas and particle pressure, and the back experiences only gas pressure, so that the net deflection of the diaphragm reflects the pressure exerted on the

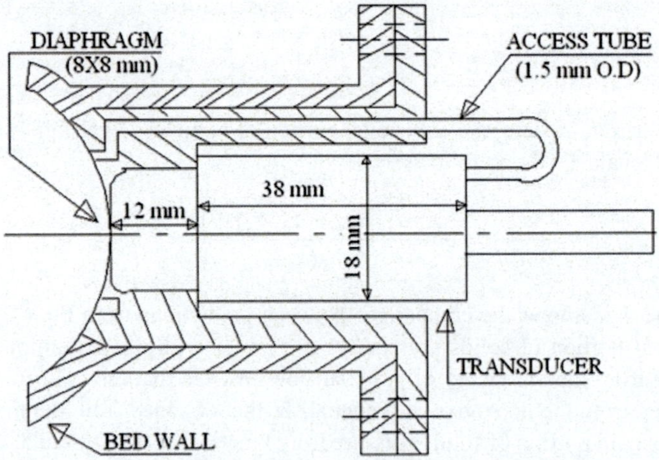


**Fig. 4.26** Particle pressure transducer

surface by particle interactions. Since in fact the gas-pressure fluctuations take a finite amount of time to reach the rear of the diaphragm, the instantaneous pressure detected by the transducer might be different from the actual solid pressure. However, its time-averaged response should correspond to the time-averaged solid pressure. In addition, a test was performed to verify that the gas pressure can be canceled out, at least on a time-averaged basis. Thus, the signal from the pressure transducer is averaged over long time periods to yield the average particle pressure.

Preliminary experiments showed that the placement of the gas ports surrounding the diaphragm could be critical. In a quiescent fluid this is not a problem because the fluid pressure is uniform across the face of the diaphragm. Difficulties arise if there is a pressure gradient tangential to the diaphragm surface. In this case, the fluid could tend to enter a port where the pressure is large and leave through another port where the pressure is lower. This could present a particular problem when the pressure drop is large so that the path through the transducer is easier than that on the outside. To avoid this difficulty, our pressure transducer has only two fluid-pressure input ports, one located on either side of the diaphragm along a diameter oriented perpendicular to the pressure gradient. These input ports were covered with a thin membrane that had 10 holes of 40 μm diameter at each side. In this way, the pressure difference between the ports is zero. No flow is induced across the interior of the transducer; also no particles entered the rear tube of the transducer.

Typical results of variation of particle pressure with time are shown in Fig. 4.27. We determined the influence of the sampling frequency on the time-average particle pressure. At 1500 Hz there was no influence.

The influence of solid mass flux on time-average solid pressure is shown in Fig. 4.28. Solid pressure increases with increasing solid mass flux, as expected. The solids pressure increases monotonically with solids volume fraction in the range of measurements.

The distribution of time average solids pressure is shown in Fig. 4.29 as a function of solids volume fractions. The solids pressure increases monotonically with solids volume fraction in the range of measurements.

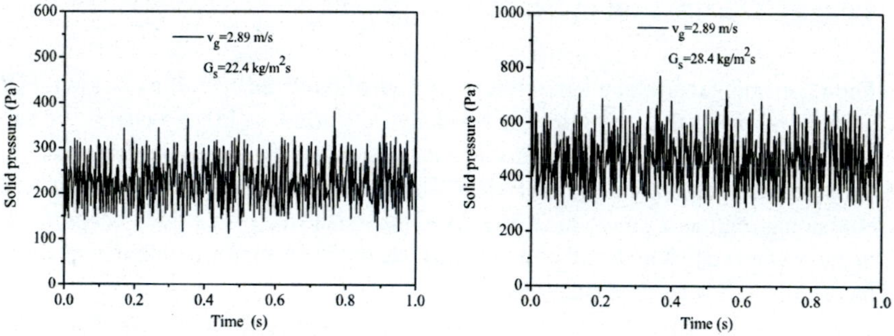


**Fig. 4.27** Instantaneous particles pressure at two different solids mass fluxes

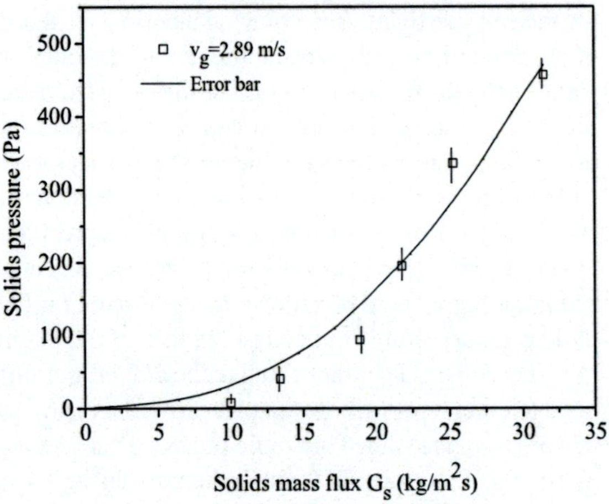


**Fig. 4.28** Effect of solid mass flux on time-averaged solids pressure

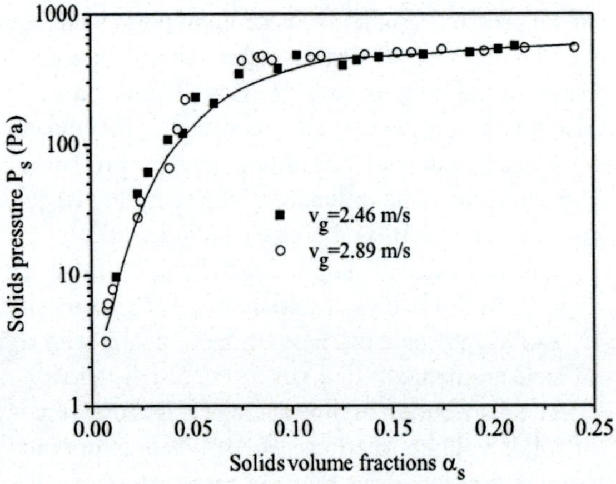


**Fig. 4.29** Time averaged solids pressure as a function of solids volume fractions

## 4.6 FCC Equation of State

The experimental granular temperature for flow of fluidized catalytic cracking (FCC) particles showed that the granular temperature (turbulent kinetic energy of particles) increases as the particle concentration increases, similar to the rise of thermal temperature upon compression of a gas. The decrease of granular temperature in the collisional regime is due to the decrease of the mean free path, which becomes zero in the packed state. On the other hand, the measured particles pressure increases with increasing solids volume fractions.

Figure 4.30 presents the equation of state. It shows that the kinetic theory of granular flow is valid for flow of FCC particles below about 5%. For dilute flow we have an analogy of the ideal gas law (Gidaspow and Huilin 1998). The ratio of solids pressure to bulk density multiplied by the granular temperature is one, as a

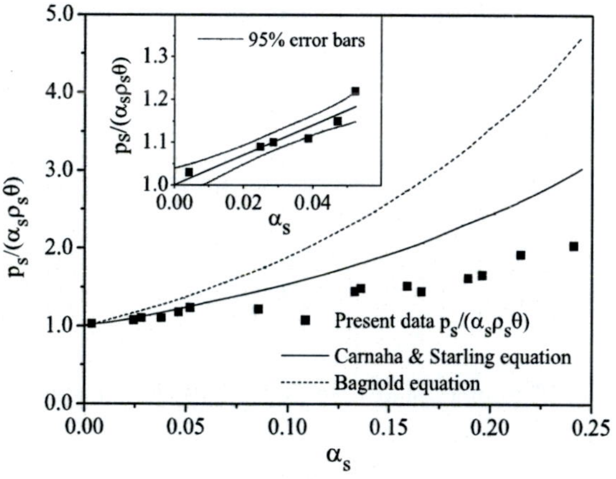


**Fig. 4.30** Equation of state for 75 μm FCC particles

limit. For volume fractions above about 5% the standard granular flow theory had to be corrected for a cohesive pressure, obtained from measurements of the radial distribution functions and granular temperature, using a modified Boltzmann relation.

From measured radial distribution function $g(r)$, we see that the collisional pressures calculated from the derivative of potential $W$ that is obtained from the first peak are the same as solid pressure collisional portion using kinetic theory of granular flow. Hence we can use the second method for further analysis. Figure 4.31 shows the potential $W$ $(r = r_o, \alpha_s)$ as a function of solids volume fraction. The derivative of this Helmholtz-type potential at a constant granular temperature is an excess or cohesion pressure, defined as

$$\frac{\partial W(r = r_0, \alpha_s)}{\partial(\alpha_s \rho_s)^{-1}} = p_{coh} \tag{4.22}$$

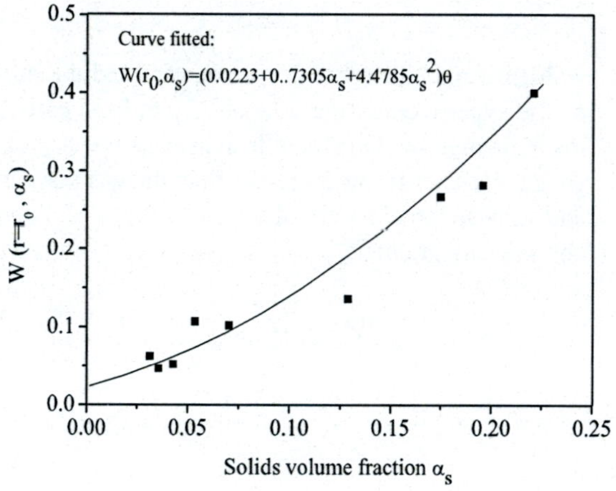


**Fig. 4.31** Variation of potential $W(r_0)$ with solid-volume fractions

It is related to the slope of the curve in Fig. 4.31 by

$$p_{coh} = -\alpha_s^2 \rho_s \frac{d[W(r = r_0, \alpha_s)]}{d\alpha_s} \tag{4.23}$$

The solids pressure then becomes:

$$p_s = p_k + p_{co} + p_{coh} = \rho_s \alpha_s \left[1 + 2(1+e)\alpha_s \rho_s - \varepsilon_s \frac{d[W(r = r_0, \alpha_s)/\theta]}{d\alpha_s}\right]\theta \tag{4.24}$$

The equation of state for FCC particles becomes as follows:

$$\frac{p_s}{\rho_s \alpha_s \theta} = 1 + 2(1+e)\alpha_s \rho_s - (0.73\alpha_s + 8.957\alpha_s^2) \tag{4.25}$$

Figure 4.30 shows the comparison between the predicted dimensionless solids pressure and the experimental data as a function of solids volume fraction. We see that the new equation of state derived based on the measured radial distribution function for FCC particles agrees with the independently measured particle pressure.

## 4.7 Solids Viscosity Measurements

For circulating fluidized bed the mixture momentum balance may be written by Tsuo and Gidaspow (1990) as,

$$\frac{\partial\left(\alpha_s \rho_s v_s^2 + \alpha_g \rho_g v_g^2\right)}{\partial z} = -\frac{\partial p}{\partial z} + \frac{1}{r}\frac{\partial}{\partial r}\left(r\left(\mu_s \frac{\partial v_s}{\partial r} + \mu_g \frac{\partial v_g}{\partial r}\right)\right) - \left(\alpha_s \rho_s + \alpha_g \rho_g\right)g \tag{4.26}$$

Miller and Gidaspow (1992) represented the mixture momentum balance based on the experiment of the flow of 75 μm FCC particles. The assumptions have been made, including (1) the solids density is much greater than the gas density. (2) the solids viscosity is much greater than the gas viscosity. (3) the radial solids velocity and velocity gradient are of the same order of magnitude as the radial gas velocity and velocity gradient

$$\frac{\partial\left(\alpha_s \rho_s v_s^2\right)}{\partial z} = -\frac{\partial p}{\partial z} + \frac{1}{r}\frac{\partial}{\partial r}\left(r\mu_s \frac{\partial v_s}{\partial r}\right) - \alpha_s \rho_s g \tag{4.27}$$

After integration and algebraic manipulation, the following equation is obtained as

$$\frac{2}{R^2}\int_0^R \frac{\partial\left(\alpha_s\rho_s\alpha_s^2\right)}{\partial z} r dr - \frac{2}{R}\mu_s \frac{\partial v_s}{\partial r}\bigg|_{r=R} = -\frac{\partial p}{\partial z} - \frac{2}{R^2}\int_0^R \alpha_s\rho_s r dr \tag{4.28}$$

This equation states that for any position $R$, the axial acceleration plus shear rate is equal to the pressure drop minus the weight of the bed. For the developed flow, the acceleration effect is zero. The particle viscosity was calculated from the equation given below:

$$\frac{\Delta p}{\Delta z} - \frac{2}{R^2}\int_0^R \alpha_s\rho_s g r dr = \frac{2}{R}\mu_s \frac{\partial v_s}{\partial r}\bigg|_{r=R} \tag{4.29}$$

A fully developed flow momentum balance shows how the velocity, concentration and pressure drop data can be used to obtain the mixture solids viscosities. The pressure drop minus the radial integrated weight of the bed over the core equals the shear at the core-annular interface. For developed flow, the solids shear equals the mixture viscosity times the radial velocity gradient. Such computations produce the viscosities (Miller and Gidaspow 1992).

Figure 4.32 shows the FCC particle viscosity measured by three difference ways, CCD camera method (Gidaspow and Huilin 1996), Brookfield viscometer and pressure drop and shear data (Miller and Gidaspow 1992). The viscosity obtained from the kinetic theory is close to the viscosity obtained from the Brookfield viscometer and pressure drop and shear data.

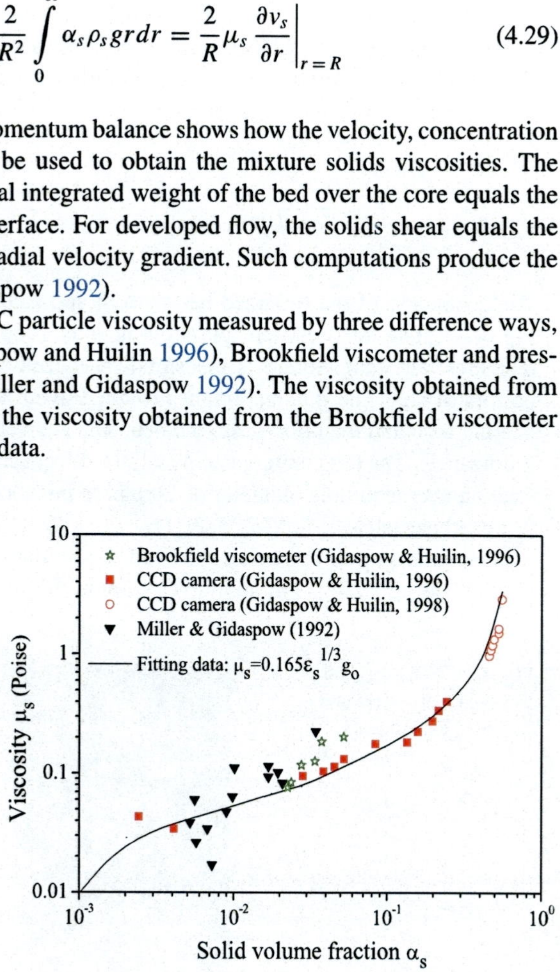


**Fig. 4.32** FCC particle viscosity measured in three different ways

## 4.8 Experiments of Extracting Alumina from Coal Fly Ash in Fluidized Bed

High-alumina coal ash is an industreferenrial byproduct which is generated in coal-fired power plants and chemical industries. Common coal ash disposal ultimately lead to the accretion of coal fly ash on wide open ground. Inappropriate management and accumulation of coal ash will be danger to both environment and human health. To utilize coal ash and decrease its damage, we need to change coal ash into high value added products. Usually, these materials are rich in $Al_2O_3$ such that the range is nearly 50% and are equivalent to mid-grade bauxite ores. Considering its high alumina content, high-alumina coal fly ash can be utilized as substitute for bauxite. Recovering alumina from coal fly ash provides a significant opportunity for converting waste materials to a new aluminum source.

### *4.8.1 Experimental Processes*

The schematic of the fluidized bed system and calcining experiments is shown in Fig. 4.33. The heat-resistant glass cylinder is 1000 mm in height and 65 mm in diameter. The temperature is measured by thermocouple which is inserted to the bottom of bed. The fluidization air provided by an air compressor flows through a heating pipe and then flows into the bed. The gas velocity is controlled by valve for fluidization. The temperature can be adjusted from 25 to 1200 °C. A steel filter plate with an averaged hole diameter of 20 μm is installed on the bottom to support the fluidized particles.

The main step of this process includes granulation, calcining, washing, crystallization and calcining. Ammonium aluminum sulfate is the key intermediate product

**Fig. 4.33** Schematic of fluidized bed system and calcining experiments

for preparation of high-purity aluminum oxide. In this process, the first step is granulation using pelletizer to prepare the bed material with ash and ammonium sulfate mixture. Ammonium sulfate reacts with alumina contained in the coal ash at high temperature, but not with the quartz and other amorphous phases. This will help to improve the purity of production after filtration. The reaction is carried out at high temperature in bubble fluidized bed. The ammonium aluminum sulfate is generated as a result of the following chemical reaction:

$$4(NH_4)_2SO_4 + Al_2O_3 \rightarrow 2NH_4Al(SO_4)_2 + 6NH_3 + 3H_2O \tag{4.30}$$

## 4.8.2 Experimental Alumina Extraction Efficiency

Figure 4.34 shows the XRD patterns of the products of the coal ash and ammonium sulfate granulated bed materials at the calcination temperature is 350 °C. The ammonium aluminum sulfate ($NH_4Al(SO_4)_2$) peak intensities gradually increase with the increase of reaction time and alumina peak intensities decrease simultaneously. This outcome reflects the fact that alumina reacted with ammonium sulfate ($(NH_4)_2SO_4$) causes the reduction of alumina peak intensities. At the same time, the quartz peak intensities keeps nearly constant or increase slightly since the amorphous quartz is generated with mullite decomposition. The ammonium aluminum sulfate peak intensities and quartz peak intensities do not change any more at reaction temperature of 350 °C, indicating that the reaction of granulated bed materials is completely performed after the calcinations time of 50 min.

The components of production of the coal ash and ammonium sulfate granulated bed materials are analyzed by XRD patterns and chemical method. For XRD method, each material has a $K$ value (or RIR value) which is calculated by $K_A = I_A / I_{Al2O3}$. The alumina extraction efficiency $\eta$ can be given as:

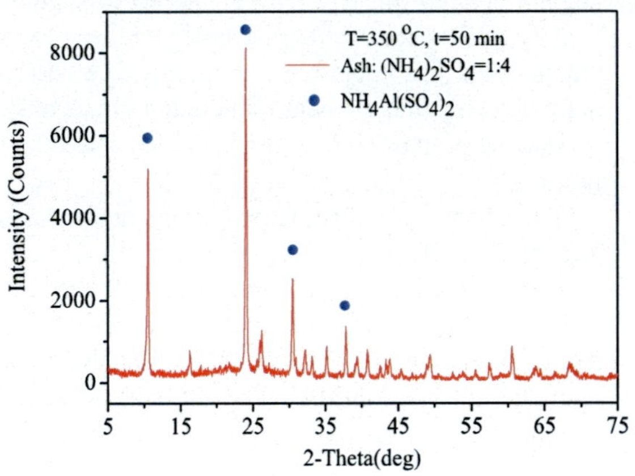


**Fig. 4.34** XRD patterns of production of granulated bed materials

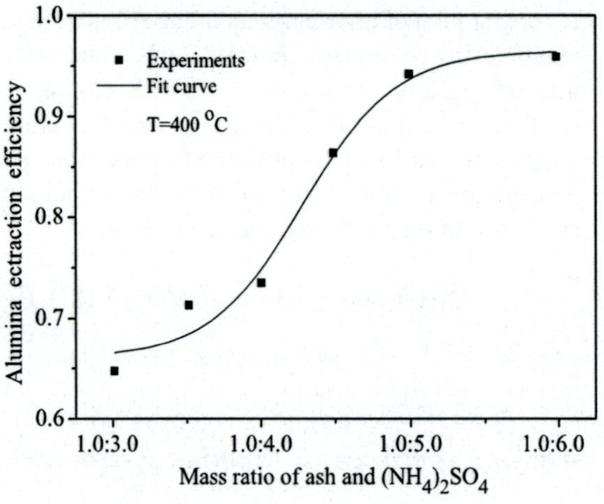


**Fig. 4.35** Profile of alumina extraction efficiency of granulated bed materials

$$\eta = \frac{M_{\mathrm{Al|NH_4Al(SO_4)_2}}}{M_{\mathrm{Al|Ash}}} \tag{4.31}$$

$$M_A = \frac{I_A}{K_A \sum_{i=1}^{N} (I_i/K_i)} \tag{4.32}$$

where $M_{\mathrm{Al|Ash}}$ and $M_{\mathrm{Al|NH_4Al(SO_4)_2}}$ are the converted alumina mass in coal fly ash and ammonium aluminum sulfate production after calcinations (Liyan et al. 2017). Figure 4.35 shows the alumina extraction efficiency of coal ash and ammonium sulfate granulated bed materials under different mass ratio between ammonium sulfate and coal ash at 400 °C at the calcination time of 50 min. The alumina extraction efficiency increases with the mass ratio between ash and ammonium sulfate. When the ratio ranges from 1.0:3.0 to 1.0:5.0, alumina extraction efficiency increases sharply. And when the mass ratio is larger than 5.0, the extraction efficiency increases slightly or keep constant. This result is attributed to the fact that the alumina in the coal fly ash has been exhausted and more feed reactants will cause no change for the reaction. For this operating condition, we can derive the liner relation between the alumina extraction efficiency and ratio of reactants for predicting reaction rate. The reasonable proportion is between 1.0:5.0 and 1.0:6.0 of ash and ammonium sulfate mixture.

The effects of calcinating temperature and calcination time on alumina extraction efficiency are depicted in Fig. 4.36. At different calcinating temperature in the bed, alumina extraction efficiency increases with the increase in calcination time from 20 to 50 min. At calcinating temperatures of 350–400 °C, alumina extraction efficiency increases sharply before 40 min and gradually at 50 min. When the results of various temperatures are compared, the alumina extraction efficiency increases fastest at reaction temperature 350 °C, but the maximum value of efficiency is at calcinating

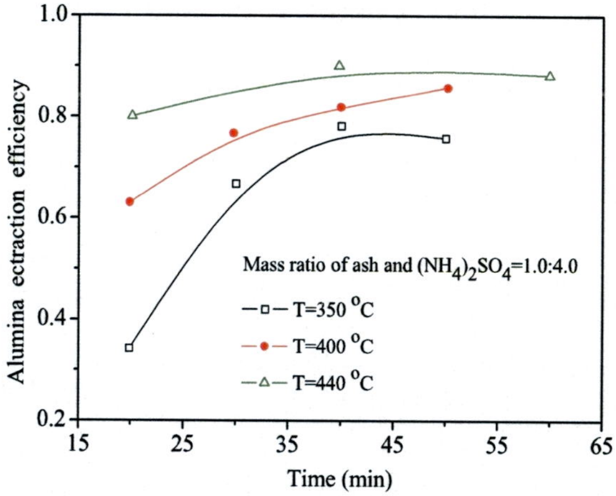


**Fig. 4.36** Profile of alumina extraction efficiency at three of granulated bed materials

temperature 440 °C. When the time needed for calcination reaction compared with hydro-chemical process and acid processes (more than 3.0 h), the present method is time saving and simple.

Figure 4.37 shows the effect of reactant ratio on alumina extraction efficiency. The value of efficiency increases with calcination time from 20 to 40 min and does not change after 40 min. When the results of various reactants ratio are compared, the tendency of curve is similar and the maximum of alumina extraction efficiency is obtained under ratio 1.0:4.5. This result is attributed to the fact that the reactant is sufficient to drive reaction completely. The result under ratio 1.0:3.0 decreases slightly after 40 min. This may be caused by the mass loss due to the fraction and collision between particles for low strength.

The high temperature fluidized bed is used to extract alumina from coal fly ash in solid state by adding ammonium sulfate. The first step is to generate ammonium

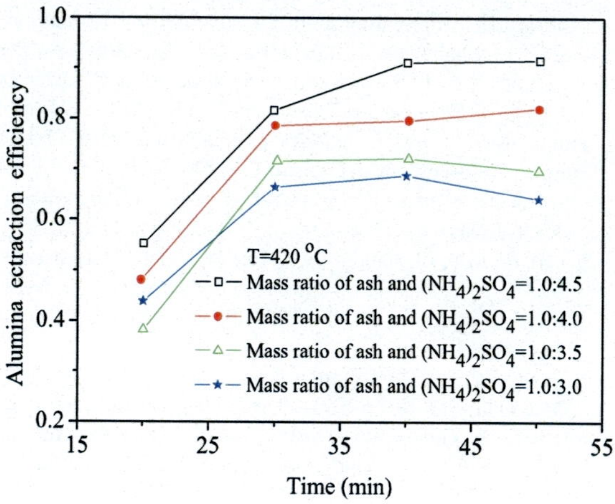


**Fig. 4.37** Profile of alumina extraction efficiency of granulated bed materials

aluminum sulfate from coal ash. Forty minutes are needed for the reaction, and the alumina extraction efficiency increases with the increase of reactant mass and the reasonable ratio is about 5.0:1.0. The maximum of extracting efficiency reaches nearly 90% at high temperature under laboratory conditions. The possibility of scale-up and improvement of the technology will be investigated.

## References

Bagnold, R. A. (1954). Experiments on a gravity free dispersion of large solid spheres in a Newtonian fluid under shear. *Proceedings of the Royal Society of London, 225,* 49–63.

Balucani, U., & Zoppi, M. (1994). *Dynamics of the liquid state.* Oxford: Clarendon Press, New York: Oxford University Press.

Carnahan, N. F., & Starling, K. E. (1969). Equation of state for Nonattracting rigid spheres. *Journal of Chemical Physics, 51,* 635–636.

Egelstaff, P. A. (1967). *An Introduction to the Liquid State.* New York: Academic Press.

Gidaspow, D. (1994). *Multiphase flow and fluidization: Continuum and kinetic theory descriptions.* California: Academic Press.

Gidaspow, D., & Huilin, L. (1996). Collisional viscosity of FCC particles in a CFB. *AICHE Journal, 42,* 2503–2510.

Gidaspow, D., & Huilin, L. (1998). Equation of state and radial distribution functions of FCC particles in a CFB. *AICHE Journal, 44,* 279–291.

Gidaspow, D., Huilin, L., & Mostofi, R. (2001). Large scale oscillations or gravity waves in risers and bubbling beds. Fluidization. In X. M. Kwauk, J. Li, W. –C. Yang (Eds.) (pp. 317–323). New York.

Gidaspow, D., Lin, C., & Seo, Y. (1983). Fluidization in two-dimensional beds with a jet. 1. Experimental porosity. *Industrial and Engineering Chemistry Research, 22,* 187–193.

Gidaspow, D., Jung, J., & Singh, R. K. (2004). Hydrodynamics of Fluidization using Kinetic Theory: An Emerging Paradigm 2002 Flour-Daniel Lecture. *Powder Technology, 148,* 123–141.

Gidaspow, D., & Mostofi, R. (2003). Maximum carrying capacity and granular temperature of A, B and C particles. *AICHE Journal, 49,* 831–843.

Hansen, J. P., & McDonald, I. R. (1991). *Theory of simple liquids.* San Diego: Academic Press.

Huilin, L. (2017). Kinetic theory for dense fluid-solid flow. Science Press.

Huilin, L., Gidaspow, D., & Bouillard, J. X. (1997). Dimension measurements of hydrodynamic attractors in circulating fluidized beds. *Powder Technology, 90,* 179–185.

Huilin, L., Shuyan, W., Yunhua, Z., Yang, L., Gidaspow, D., & Jianmin, D. (2005). Prediction of particle motion in a two-dimensional bubbling fluidized bed using discrete hard-sphere model. *Chemical Engineering Science, 60,* 3217–3231.

Hunter, R. J. (1989). *Foundation of colloid science II.* Oxford: Oxford Science Publications.

Jung, J., Gidaspow, D., & Gamwo, I. K. (2005). Measurement of two kinds of granular temperatures, stresses and dispersion in bubbling beds. *Industrial and Engineering Chemistry Research, 44,* 1329–1341.

Liyan, S., Kun, L., Jianren, F., & Huilin, L. (2017). Experimental study of extracting alumina from coal fly ash using fluidized beds at high temperature. *Fuel, 199,* 22–27.

Miller, A., & Gidaspow, D. (1992). Dense vertical gas-solid flow in a pipe. *AICHE Journal, 38,* 1801–1815.

Savage, S. B. (1983). Granular flows at high shear rates. In R. E. Meyer (Ed.), *Theory of Dispersed Multiphase Flow* (pp. 339–358). New York: Academic Press.

Seo, Y., & Gidaspow, D. (1987). An X-Ray-Gamma-ray of measurement of binary solids concentrations and void in fluidized beds. *Industrial and Engineering Chemistry Research, 26,* 1622–1632.

Shuyan, W., Guodong, L., Huilin, L., Yinghua, B., Jianmin, D., & Yunha, Z. (2009). Prediction of radial distribution function of particles in a gas-solid fluidized bed using discrete hard-sphere model. *Industrial and Engineering Chemistry Research, 48,* 1343–1352.

Shuai, W. (2014). Large eddy approach and meso-macro scale effect on dense gas-solid flow. Ph.D. thesis, Harbin Institute of Technology, Harbin.

Tsuo, Y. P., & Gidaspow, D. (1990). Computation of flow patterns in circulating fluidized beds. *AICHE Journal, 36,* 885–896.

# Chapter 5
# Tutorial for Numerical Methods and Program Solution Technique

## Nomenclature

$D$ Diffusion coefficient, $m^2/s$
$h$ Enthalpy, kJ/kg
$p$ Pressure, $N/m^2$
$t$ Time, s
$T$ Temperature, K
$u$ Velocity, m/s
$v$ Velocity, m/s
$x$ Distance, m

## Greek Symbols

$\alpha$ Volume fraction
$\beta$ Interphase friction coefficient, $kg/m^3s$
$\tau$ Stress, $N/m^2$
$\rho$ Density, $kg/m^3$
$\theta$ Granular temperature, $m^2/s^2$

## Subscripts

$f$ Fluid phase
$k$ Phase $k$
$s$ Solids phase



H. Lu et al., *Computational Fluid Dynamics and the Theory of Fluidization*,
https://doi.org/10.1007/978-981-16-1558-0_5

## 5.1 Conservation Laws and Convection Diffusion

Conservation equations for mass, momentum and energy for each component and each phase in a mixture can be obtained using the Reynolds transport theorem (Gidaspow 1994; Aris 1962). As an example, consider conservation of mass of species *i*, with rate of consumption of species per unit volume. The mass balance for species *i* for a system of constant mass, $m_i$ moving with velocity $\vec{v}_i$ simply states that the rate of change of $m_i$ is its rate of production. In one dimension the primitive form obtained from a conservation law is as follows, where the subscript "*i*" standing for *i* species has been dropped for convenience.

$$\frac{\partial(v\rho)}{\partial x} = 0 \tag{5.1}$$

Equation (5.1) is the steady one dimensional conservation of mass equation, which when integrated states that the mass flux per unit area is a constant.

The derivative in Eq. (5.1) can be finite differenced using the second order central finite difference representation, where I is associated with the variable *x*.

$$\frac{\partial(\rho v)}{\partial x} = \frac{(\rho v)(I+1) - (\rho v)(I-1)}{2\Delta x} + error(I) \tag{5.2}$$

We want to compute the error in the mass flux in *N* steps.

$$\int_{x=0}^{x=N\Delta x} \frac{\partial(\rho v)}{\partial x} dx = \frac{1}{2}\sum_{i=1}^{N}[(\rho v)(I+1) - (\rho v)(I-1)] \tag{5.3}$$

The negative of the summation in Eq. (5.3) is as follows

$$\sum_{i=1}^{N}[(\rho v)(I-1) - (\rho v)(I+1)] = (\rho v)(0) + (\rho v)(1) - (\rho v)(N) - (\rho v)(N+1) \tag{5.4}$$

Hence the integral of the flux becomes as follows

$$\int_{x=0}^{x=N\Delta x} \frac{\partial(\rho v)}{\partial x} dx = \frac{1}{2}[(\rho v)(N) + (\rho v)(N+1)] - \frac{1}{2}[(\rho v)(0) + (\rho v)(1)]$$

$$= [(\rho v)]\left(N + \frac{1}{2}\right) - [(\rho v)]\left(\frac{1}{2}\right) \tag{5.5}$$

The intermediate values have canceled out, except for the possible round off error. The net flux is the flux out at the cell face $N + \frac{1}{2}$ minus the flux at face $\frac{1}{2}$. The mass is conserved.

### *5.1.1 Nonconservative Form*

Let us now finite difference a typical convective term in Eq. (5.1)

$$v\frac{\partial \omega}{\partial x} \tag{5.6}$$

where the subscript $i$ on the weight fraction $\omega$ has been dropped for convenience. Using the same second order finite difference expression, we obtain

$$v\frac{\partial \omega}{\partial x} = v(I)\frac{\omega(I+1) - \omega(I-1)}{2\Delta x} + error(I) \tag{5.7}$$

The net volumetric flux from I + 1 to I–1 according to Eq. (5.7) is

$$\text{volumetric flux} = \mathrm{v}(I)\omega(I+1) - v(I)\omega(I-1) \tag{5.8}$$

The true volumetric flux is obtained from the conservative form of the equation

$$\frac{\partial(\omega\rho)}{\partial x} = 0 \tag{5.9}$$

as

$$\text{volumetric flux} = (v\omega)(I+1) - (v\omega)(I-1) \tag{5.10}$$

The error made at each step is

$$\text{error at step I} = [v(\mathrm{I}+1) - v(I)]\omega(I+1) + [v(I) - v(I-1)]\omega(I-1) \tag{5.11}$$

The error in $N$ steps is as follows,

$$\text{error in N steps} = \sum_{\mathrm{i}=1}^{\mathrm{N}} \{[v(\mathrm{I}+1) - v(I)]\omega(I+1) + [v(I) - v(I-1)]\omega(I-1)\} \tag{5.12}$$

The error accumulates as the grid is refined, that is as $N$ becomes large. The error is due to the change in the velocity. The error can be canceled by the use of a transportive conservative representation of the convective term in Eq. (5.6).

### 5.1.2 Conservative Upwind (Donor Cell) Form

The loss of mass error can be eliminated by several means. The simplest is to use an upwind or backward difference for the derivative and to average the velocity.

$$v\frac{\partial\omega}{\partial x}=\begin{cases}\frac{v(I+1)+v(I)}{2\Delta x}\omega(I)-\frac{v(I+1)+v(I-1)}{2\Delta x}\omega(I-1)\ for & \begin{matrix}v(I)>0\\ v(I+1)>0\end{matrix}\\ \frac{v(I+1)+v(I)}{2\Delta x}\omega(I+1)-\frac{v(I)+v(I-1)}{2\Delta x}\omega(I)\quad for & \begin{matrix}v(I)\le 0\\ v(I+1)\le 0\end{matrix}\end{cases} \tag{5.13}$$

The justification for the use of backward differences will be given in the next section. Here we see that the conservation of mass error does not occur in this algorithm. For positive velocities, the volumetric flux is as follows.

$$\int_{i=I}^{i=N} v\frac{\partial\omega}{\partial x}dx=\frac{v(N+1)+v(N)}{2}\omega(N)-\frac{v(I)+v(I-1)}{2}\omega(I-1) \tag{5.14}$$

This upwind or "donor" cell method (Harlow and Amsden 1971) has a second order accuracy in space derivatives. For constant $\omega$ and positive velocities, Eq. (5.13) gives the expression

$$\frac{\partial(v\omega)}{\partial x}=\frac{\omega[v(I+1)+v(I)-v(I+1)-v(I-1)]}{2\Delta x}=\omega\frac{v(I)-v(I-1)}{2\Delta x} \tag{5.15}$$

The right hand side of Eq. (5.15) has an accuracy of $(\Delta x)^2$.

### 5.1.3 Other Conservative Forms

Harlow and Amsden (1971) proposed a method for conservative differencing of flux expressions, in addition to the donor cell method. In their technique the scalar variables, such as the density, the internal energy and the volume fraction are located at the center of the cell, at location (I, J) in two dimensions. The velocities are evaluated at the cell edges, e.g. at (I + 1/2, J + 1/2), seeing in Fig. 5.1.

Using such a notation, the derivative of the mass flux is written as,

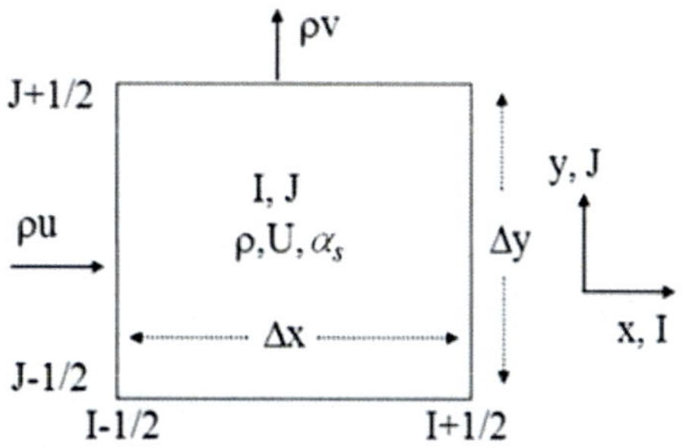


**Fig. 5.1** Computational cell in the ICE method for mass and energy balances

$$\frac{\partial(\rho u)}{\partial x} = \frac{(\rho u)(I+1/2) - (\rho u)(I-1/2)}{\Delta x} + O(\Delta x) \tag{5.16}$$

In the donor cell method we already had

$$(\rho u)(I+1/2) = \begin{cases} \rho(I)[u(I)+u(I+1)]/2 & if\ \ u(I)+u(I+1) > 0 \\ \rho(I+1)[u(I)+u(I+1)]/2 & if\ \ u(I)+u(I+1) < 0 \end{cases} \tag{5.17}$$

### *5.1.4 Upwinding*

Upwind finite difference schemes, refereed to simply as upwinding, are based on the theory of first order partial differential equations (Courant and Hilbert 1962; Gidaspow 1994). This theory shows that a hyperbolic system of first order partial differential equations can be transformed into a system of ordinary differential equations, along the characteristic paths. Information flows along these paths. To obtain a solution in a given region, information must flow into this region. If all characteristics are positive, inlet conditions are sufficient to determine a unique solution. If one characteristic is negative, a boundary condition must be given. This means information must be given "upwind' to obtain a unique solution. This concept is best illustrated by an example, say the familiar shower problem.

When a pump sends hot water from the central heating system to users, we commonly experience cold water running out of the transporting pipe system for a few minutes until hot water from the heatingcenter has reached the user's locations. This is due to the presence of cold water in the pipe between the supply hot heater and the heating users. An adiabatic energy balance for water in the pipe is as follows.

$$\frac{d}{dt}\int_{x}^{x+\Delta x} ah\rho dx + ah\rho v\,]_{x}^{x+\Delta x} = 0 \tag{5.18}$$

When $h = h(T)$ only, the application of the mean value theorems (Gidaspow 1994) gives the differential equation for constant area "$a$" as follows.

$$\frac{\partial T}{\partial t} + v\frac{\partial T}{\partial x} = 0 \tag{5.19}$$

where $T$ is the water temperature, $t$ is time, $x$ is the length of pipe and $v$ is the velocity of water.

By inspection or by the method of characteristics (Gidaspow 1994), Eq. (5.19) is equivalent to the first order ordinary differential equation

$$\frac{dT}{dt} = 0 \tag{5.20}$$

along the path

$$\frac{dx}{dt} = v \tag{5.21}$$

Equations (5.20) and (5.21) show that the temperature does not change with time moving with velocity $v$. For the supplying hot water using a pump we had the initial condition for the water in the pipe;

$$\mathrm{T} = \mathrm{T_{cold}}(\mathrm{x}) \quad \text{at} \quad \mathrm{t} = 0 \tag{5.22}$$

where $T_{cold}$ was a function of the distance $x$. When the pump was turned on, hot water started flowing into the pipe. The inlet condition, at $x = 0$ was

$$\mathrm{T} = \mathrm{T_{hot}}(\mathrm{t}) \quad \text{at} \quad \mathrm{x} = 0 \tag{5.23}$$

where $T_{hot}$ was a function of time. Figure 5.2 shows the analytical solution for this problem. The figure shows that up to time $= L/v$ cold water of temperature $T(x)$

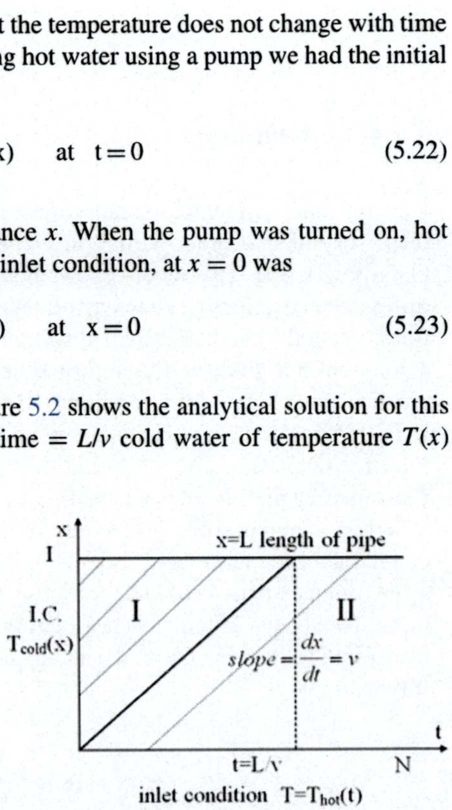


**Fig. 5.2** Temperature variations with times

exits the pipe system. After that time, hot water starts flowing. The inaccuracy in the physics is that heat transfer from the water to the pipe wall was deleted in the formulation.

In the donor cell method, $v\frac{\partial T}{\partial x}$ is finite differenced as follows.

$$v\frac{\partial T}{\partial x} = v\frac{T(N,\ J) - T(N,\ J-1)}{\Delta x} \qquad if \quad v > 0 \tag{5.24}$$

and

$$v\frac{\partial T}{\partial x} = v\frac{T(N,\ J+1) - T(N,\ J)}{\Delta x} \qquad if \quad v < 0 \tag{5.25}$$

Patankar (1980) calls this UPWIND finite differencing. The fluid enters the cell J with a temperature at its upwind condition. As shown in Fig. 5.3, the fluid enters the cell J with a temperature at its upwind condition.Using the donor celllogic, the finite difference representation of Eq. (5.19) becomes as follows for $v > 0$.

$$\frac{T(N+1,\ J) - T(N,\ J)}{\Delta t} + v\frac{T(N,\ J) - T(N,\ J-1)}{\Delta x} = 0 \tag{5.26}$$

In Eq. (5.26) the convective term was finite differenced explicitly, at time $N$. It is often IMPLICITELY finite differenced at time, $N+\ 1$. The EXPLICIT scheme (5.26) gives:

$$T(N+1,\ J) = T(N,\ J)\left[1 - \frac{v\Delta t}{\Delta x}\right] + v\frac{\Delta t}{\Delta x}T(N,\ J-1) \tag{5.27}$$

Let

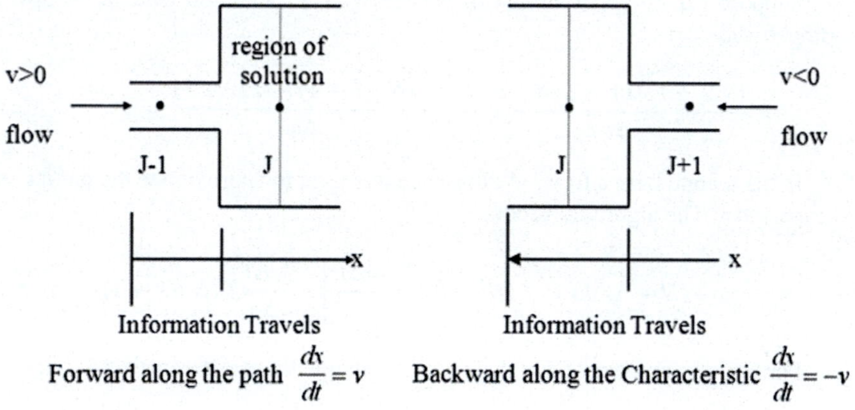


**Fig. 5.3** Tube-tank upwind scheme

$$\Delta t = \frac{\Delta x}{v} \tag{5.28}$$

This amounts to integration along the characteristic, $dx/dt = v$.

In region I we obtain all zeros, along the line $x = vt$. In region II we obtain a constant temperature. Finite difference representation Eq. (5.27) gives the exact solution to the differential equation, since Eq. (5.19) states that

$$\frac{dT}{dt} = 0 \tag{5.29}$$

along $dx/dt = v$.

Suppose in Eq. (5.27) we let

$$\Delta t > \frac{\Delta x}{v} \tag{5.30}$$

For the case

$$\Delta t = \frac{3\Delta x}{v} \tag{5.31}$$

The problem was that information from the initial condition was used to find values determined by the inlet condition. To prevent this, we must let

$$\Delta t \le \frac{\Delta x}{v} \text{ (Courant Condition)} \tag{5.32}$$

The proof can be obtained also from the von Neumann stability analysis. More simply, it follows from positive coefficient test. To prevent negative values of temperature, Eq. (5.27) shows that the criterion given by Eq. (5.32) must be satisfied.

Suppose Eq. (5.27) is finite-differenced with no attention paid to its upwind property, as

$$\frac{T(N+1,\ J) - T(N,\ J)}{\Delta t} + v\frac{T(N,\ J+1) - T(N,\ J)}{\Delta x} = 0 \tag{5.33}$$

In the second term a forward difference was used to approximate the partial with respect to $x$. The algorithm becomes

$$T(N+1,\ J) = T(N,\ J)\left[1 + \frac{v\Delta t}{\Delta x}\right] - \frac{v\Delta t}{\Delta x}T(N,\ J+1) \tag{5.34}$$

For Courant number of one this gives the simple finite difference equation

$$T(N+1,\ J) = 2T(N,\ J) - T(N,\ J+1) \tag{5.35}$$

This algorithm is inconsistent with the analytical solution to

$$\frac{dT}{dt} = 0 \qquad \text{along} \qquad \frac{dx}{dt} = v \tag{5.36}$$

which in finite difference form of the same order as Eq. (5.33) is

$$T(N+1,\ J) = T(N,\ J) \tag{5.37}$$

where $\Delta x = v\Delta t$.

Equation (5.34) is also unstable and violates the positive coefficient test. However, its disagreement with the analytical solution to the partial differential equation is the strongest test of its incorrectness. This test can be applied to some high order upwind methods that have been proposed.

## 5.2 Numerical Diffusion

Consider the same differential linear operator, $L$

$$L(T) = \frac{\partial T}{\partial t} + v\frac{\partial T}{\partial x} = 0 \tag{5.38}$$

donor cell differenced, $L_\Delta$ as

$$L_\Delta(T) = \frac{T(N+1,\ J) - T(N,\ J)}{\Delta t} + v\frac{T(N,\ J) - T(N,\ J-1)}{\Delta x} = 0 \tag{5.39}$$

By Taylor series,

$$T(N+1,\ J) = T(N,\ J) + \frac{\partial T}{\partial t}\cdot \Delta t + \frac{1}{2}\left.\frac{\partial^2 T}{\partial t^2}\right|_{mean} \cdot (\Delta t)^2 \tag{5.40}$$

and

$$T(N,\ J-1) = T(N,\ J) - \frac{\partial T}{\partial x}\cdot \Delta x + \frac{1}{2}\left.\frac{\partial^2 T}{\partial x^2}\right|_{mean} \cdot (\Delta x)^2 \tag{5.41}$$

Hence the truncation error, $\tau$ is

$$\tau = L(T) - L_\Delta(T) = -\frac{1}{2}\left.\frac{\partial^2 T}{\partial t^2}\right|_{mean}(\Delta t) + \frac{1}{2}\left.\frac{\partial^2 T}{\partial x^2}\right|_{mean}(\Delta x)\cdot v \tag{5.42}$$

But since

$$\frac{dx}{dt} = v, \frac{\Delta x}{\Delta t} = v \tag{5.43}$$

$$\tau = \frac{1}{2}\Delta t\left(\frac{\partial^2 T}{\partial t^2} - v^2\frac{\partial^2 T}{\partial x^2}\right) \tag{5.44}$$

The truncation error satisfies a wave equation, Eq. (5.44) to zero order in $\Delta t$.

$$\frac{\partial^2 T}{\partial t^2} = v^2\frac{\partial^2 T}{\partial x^2} \tag{5.45}$$

Then the truncation error may be written as

$$\tau = \frac{1}{2}v\Delta x\left(1 - \frac{v\Delta t}{\Delta x}\right)\frac{\partial^2 T}{\partial t^2} \tag{5.46}$$

where we obtain a numerical diffusion coefficient,

$$D_{dif} = \frac{1}{2}v\Delta x\left(1 - \frac{v\Delta t}{\Delta x}\right) \tag{5.47}$$

Then we see that by solving $L_\Delta = 0$, we have actually obtained a solution to

$$\frac{\partial T}{\partial t} + v\frac{\partial T}{\partial x} = D_{dif}\frac{\partial^2 T}{\partial x^2} \tag{5.48}$$

a convective diffusion equation rather than Eq. (5.48).

Note that along a characteristic

$$\frac{dx}{dt} = v \tag{5.49}$$

$$D_{dif} = 0 \tag{5.50}$$

Also note that for a diffusion coefficient to be positive, a requirement of the second law of thermodynamics

$$\frac{v\Delta t}{\Delta x} \le 1 \tag{5.51}$$

The Courant condition is thus obtained.

## 5.3 Optimum Time Step

For a finite difference solution of a wave equation the optimum accuracy is achieved by taking the time step equal to the transit time, as determined by the characteristic directions. In other words optimum accuracy is achieved by setting the Courant number to one. Equation (5.47) shows that for Courant number of one, that is for

$$\Delta t = \Delta x / v \tag{5.52}$$

Numerical diffusion $D_{num}$ vanishes.

$$D_{num} = 0 \tag{5.53}$$

In practice, where $v$ changes, as the computation proceeds one needs to change the time step. Indeed this has to be done in multiphase flow calculations. When the gradients are steep, one takes a very small $\Delta t$. After the shock has passed one increases $\Delta t$ to a reasonable value. This approach enables computations to be done in a reasonable time. At the same time, numerical diffusion is minimized.

Equation (5.47) further shows that for multiphase flow in vertical pipes, where the flow is close to developed, the numerical diffusion in the radial direction is very small, since the radial velocity component is also very small.

$$D_{radial} = \frac{1}{2} v_{radial} \cdot \Delta y \left(1 - \frac{v_{radial} \Delta t}{\Delta x}\right) \tag{5.54}$$

In the direction of flow the convective terms dominate the solution. Both the numerical and the physical diffusion are unimportant compared with convection. Hence numerical diffusion is not a big issue in multiphase flow computations using the upwind method.

## 5.4 Flux-Corrected Transport

The flux corrected transport (FCT) algorithm is used to compute steep gradients and inviscid shocks. This method minimizes the numerical diffusion.

The temperature is computed by advancing in time one step from $N$ to $N + 1$. In space one uses a three-point explicit finite difference formula to obtain greater accuracy. One writes

$$T(N+1, J) = a_j T(N, J-1) + b_j T(N, J) + c_j T(N, J+1) \tag{5.55}$$

Equation (5.27) is a special case of Eq. (5.55). Energy conservation is preserved by requiring

$$a_j + b_j + c_j = 1 \tag{5.56}$$

To obtain positive temperature one also requires that all $a_i$, $b_i$ and $c_i$ be positive. In the end one obtains a form

$$T(N+1, J) = \mu\left(J - \frac{1}{2}\right)[T(J) - T(J-1)] - \mu\left(J + \frac{1}{2}\right)[T(J+1) - T(J)] \tag{5.57}$$

where $\mu(J + \frac{1}{2})$ is an antidiffusion coefficient.

## 5.5 Staggered Mesh Differencing and the ICE Method

In the finite volume method a region in space is divided into a set of control volumes. The conservation laws are applied to each control volume. Hence with a correct representation of a control volume and surface, the conservation laws are automatically satisfied. There are two broad categories of meshes used in finite volume methods: structured and unstructured finite volume meshes. They are represented in Fig. 5.4. The use of the unstructured mesh is for a better representation of arbitrary geometry than can be done with structured meshes. Since these techniques have not been perfected as yet, for the solution of Navier–Stokes' equations and multiphase problems, further discussion will be restricted to the ICE method and related structured mesh methods.

Almost all two fluid codes used the staggered mesh. Spalding's SIMPLE (semi-implicit method for pressure-linked equations) algorithm for solving the Navier–Stokes equations also uses this concept (Anderson 1995). The staggered grid concept is needed for convergence on pressure. If the pressure gradient is finite differenced by second order differences, then it is possible to obtain convergence with a non-uniform pressure fluid. The ICE (implicit continuous Eulerian) technique has been used for solving Navier–Stokes equations (Harlow and Amsden 1971; Kashiwa et al. 1994).

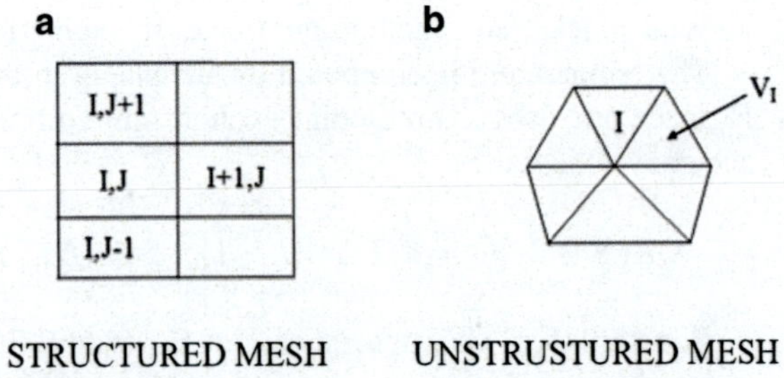


**Fig. 5.4** Structured (**a**) and unstructured (**b**) finite volume meshes

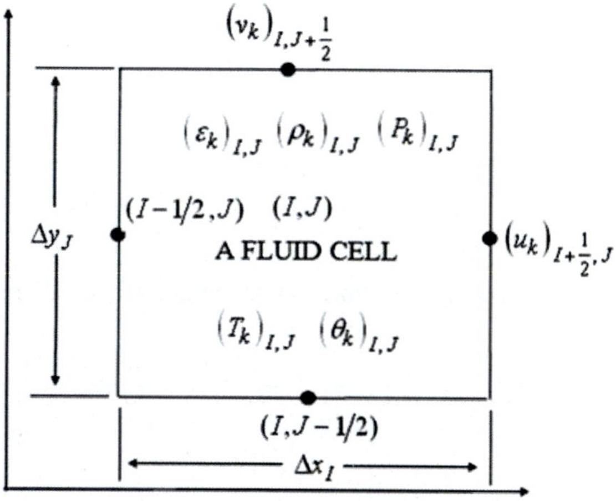


**Fig. 5.5** Computational mesh for mass, energy and granular temperature balances for phase *k*

## 5.5.1 Computational Grid

The basic concept of the staggered mesh has been shown by examples. Figure 5.5 shows a computational mesh for a two dimensional region. The physical coordinates $(x, y)$ of each cell center is described by the integer pair (I, J). The position of the cell center is denoted by the distances $x_I$, $y_J$. The cell dimensions or the distances between grid lines are denoted by $\Delta x_I$ and $\Delta y_J$ in their respective directions. They can be variable. The physical location of the grid lines are denoted by distances such as $x_{I+\frac{1}{2}}$, $y_{J+\frac{1}{2}}$. The index notation $I+\frac{1}{2}$ indicates a location between cell positions I and I + 1. Similarly $I-\frac{1}{2}$ is between I–1 and I. The density of phase *k*, its volume fraction $\alpha_k$, the thermal temperature $T_k$ and the granular temperature $\theta_k$ are evaluated at the cell center, (I, J). The phase velocities components in the *x* direction, $u_k$ and in the y direction $v_k$are evaluated at the edges.

## 5.5.2 Differencing Technique for Mass and Momentum

First order differencing will be illustrated for the continuity equation in two dimensions.

$$\frac{\partial \rho}{\partial t}+\frac{\partial(\rho u)}{\partial x}+\frac{\partial(\rho v)}{\partial y}=0 \tag{5.58}$$

For convenience the standard notation for time differencing will be represented by the superscript n. Thus the finite difference representation for the partial derivative of the density

$$\frac{\partial \rho}{\partial t} = \frac{\rho(t, x, y) - \rho(t + \Delta t, x, y)}{\Delta t} + O(\Delta t) \tag{5.59}$$

will be written as

$$\frac{\partial \rho}{\partial t} = \frac{\rho^{n+1} - \rho^{n}}{\Delta t} = \left.\frac{\Delta \rho}{\Delta t}\right|_{n}^{n+1} \tag{5.60}$$

The convective fluxes are evaluated at the cell edges, as follows.

$$\frac{\partial(\rho u)}{\partial x} = \frac{(\rho u)\left(I + \frac{1}{2}, J\right) - (\rho u)\left(I - \frac{1}{2}, J\right)}{\Delta x} + O(\Delta x) \tag{5.61}$$

$$\frac{\partial(\rho v)}{\partial y} = \frac{(\rho v)\left(I + \frac{1}{2}, J\right) - (\rho v)\left(I - \frac{1}{2}, J\right)}{\Delta y} + O(\Delta y) \tag{5.62}$$

But in Eqs. (5.61) and (5.62) $\rho$ is not defined at the cell edges, e.g., at I + 1/2. Using the upwind logic we define

$$\rho\left(I + \frac{1}{2}, J\right) = \begin{cases} \rho(I, J) & if \quad u\left(i + \frac{1}{2}\right) \geq 0 \\ \rho(I+1, J)\, if & u\left(i + \frac{1}{2}\right) < 0 \end{cases} \tag{5.63}$$

The same definitions are used for the partial density $\rho_k$ , the volume fraction $\alpha_k$ and granular temperature $\theta_k$.

The momentum equation in the $x$ direction

$$\frac{\partial(\rho u)}{\partial t} + \frac{\partial(\rho u u)}{\partial x} + \frac{\partial(\rho u v)}{\partial y} = -\frac{\partial p}{\partial x} \tag{5.64}$$

is centered at $\left(I + \frac{1}{2}, J\right)$ as shown in Fig. 5.6. The momentum in the $y$ direction is centered at $\left(I, J + \frac{1}{2}\right)$.The partials in Eq. (5.64) are evaluated as follows

$$\frac{\partial(\rho u)}{\partial t} = \frac{(\rho u)^{n+1}\left(I + \frac{1}{2}, J\right) - (\rho u)^{n}\left(I + \frac{1}{2}, J\right)}{\Delta t} = \left.\frac{\Delta \rho u\left(I + \frac{1}{2}, J\right)}{\Delta t}\right|_{n}^{n+1} \tag{5.65}$$

The pressure is evaluated at the cell center

$$\frac{\partial p}{\partial x} = \frac{p(I+1, J) - p(I, J)}{\Delta x_{I+\frac{1}{2}}} + O(\Delta x) \tag{5.66}$$

where

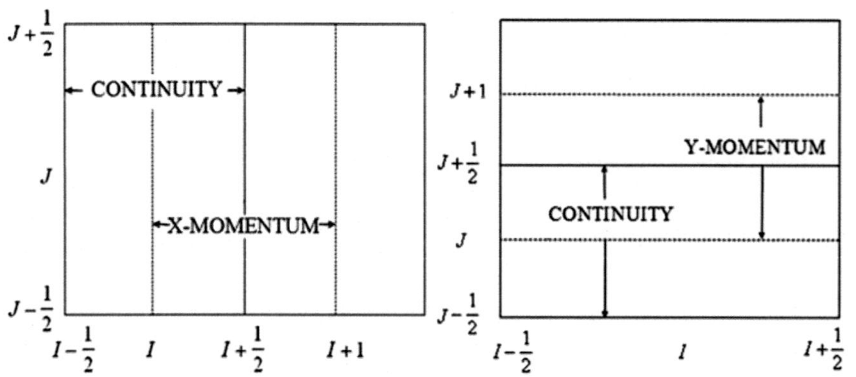


**Fig. 5.6** The staggered computational mesh for momentum equations

$$\Delta x_{I+\frac{1}{2}} = \frac{1}{2}(\Delta x_I + \Delta x_{I+1}) \tag{5.67}$$

The $x$-component momentum flux is

$$\frac{\partial(\rho uu)}{\partial x} = \frac{(\rho uu)(I+1, J) - (\rho uu)(I, J)}{\Delta x_{I+\frac{1}{2}}} \tag{5.68}$$

This derivative requires definitions of the velocity at all centers in the same way that scalars required definition at all boundaries. A combination of averaging and upwind ideas is used to make the definitions. The average mass flux at I is defined to be the average of the adjacent flows at $I - \frac{1}{2}$ and $I + \frac{1}{2}$, as

$$\rho u(I, J) = \frac{1}{2}\left[(\rho u)\left(I - \frac{1}{2}, J\right) + (\rho u)\left(I + \frac{1}{2}, J\right)\right] \tag{5.69}$$

The velocity is defined using the upwind logic

$$u(I, J) = \begin{cases} u\left(I - \frac{1}{2}, J\right) \; if \quad u \geq 0 \\ u\left(I + \frac{1}{2}, J\right) \; if \quad u < 0 \end{cases} \tag{5.70}$$

Similarly the $y$ component momentum flux in the $x$ direction is

$$\frac{\partial(\rho uv)}{\partial y} = \frac{(\rho uv)\left(I + \frac{1}{2}, J + \frac{1}{2}\right) - (\rho uv)\left(I + \frac{1}{2}, J - \frac{1}{2}\right)}{\Delta y} + O(\Delta y) \tag{5.71}$$

Define the average flux as the area weighted average

$$\rho v\left(I+\frac{1}{2}, J+\frac{1}{2}\right)=\frac{\frac{1}{2}\left[\Delta x_I(\rho v)\left(I, J+\frac{1}{2}\right)+\Delta x_I(\rho v)\left(I+1, J+\frac{1}{2}\right)\right]}{\frac{1}{2}\left[\Delta x_I+\Delta x_{I+\frac{1}{2}}\right]} \tag{5.72}$$

and the upwind velocity as

$$v\left(I+\frac{1}{2}, J+\frac{1}{2}\right)=\begin{cases} v\left(I+\frac{1}{2}, J\right) & if \quad v \geq 0 \\ v\left(I+\frac{1}{2}, J+1\right) & if \quad v<0 \end{cases} \tag{5.73}$$

The definitions for the $y$ direction momentum balances can be written similarly, as indicated in Fig. 5.6.

### 5.5.3 Overview of Solution Technique Using a Residue

The multiphase Navier–Stokes equations can be solved explicitly, implicitly or semi-implicitly. Complete explicit solution leads to a Courant time step restriction limited by the sonic velocity (Rivard and Torrey 1977).

$$\Delta t \leq \frac{\Delta x}{\text{Sonic velocity}} \tag{5.74}$$

Hence a small $\Delta t$ is required. Complete implicit solution gives a set of nonlinear algebraic equations that are too difficult to solve. Thus in the ICE method a compromise is reached. One uses a semi-implicit method. The conservation of mass equations are finite differenced implicitly, removing the time step limitation. The convective terms in the momentum equations are finite differenced explicitly, while most other terms in the momentum equations, particularly the pressure, are finite differenced implicitly. Therefore we still have the reasonable transit time limitation,

$$\Delta t \leq \frac{\Delta x}{\text{Maximum velocity}} \tag{5.75}$$

### 5.5.4 The Finite Difference Equations and Averaging Process

Quantities in the finite difference equations required at spatial locations other than where they are defined are obtained by weighted averaging.

For cell centered quantities, the cell centered properties $\Psi$ are defined at the cell center at $(i, j)$. At other locations averaging is used as follows.

$$\Psi_{i+\frac{1}{2},j} = \frac{1}{2\delta r_{i+\frac{1}{2}}}\left(\delta r_{i+1}\Psi_{i,j} + \delta r_i \Psi_{i+1,j}\right) \tag{5.76a}$$

$$\Psi_{i,j+\frac{1}{2}} = \frac{1}{2\delta z_{j+\frac{1}{2}}}\left(\delta z_{j+1}\Psi_{i,j} + \delta z_j \Psi_{i,j+1}\right) \tag{5.76b}$$

$$\begin{aligned}\Psi_{i+\frac{1}{2},j+\frac{1}{2}} = & \frac{1}{4\delta r_{i+\frac{1}{2}}\delta z_{j+\frac{1}{2}}}(\delta r_{i+1} z_{j+1}\Psi_{i,j} + \delta r_i \delta z_{j+1}\Psi_{i+1,j} + \delta r_{i+1}\delta z_j \Psi_{i,j+1} \\ & + \delta r_i \delta z_j \Psi_{i+1,j+1})\end{aligned} \tag{5.76c}$$

For boundary centered quantities, the boundary centered quantities in $x$ direction is $u$ which is defined at $i, j + \frac{1}{2}$. The averaging is as follows.

$$u_{i,j} = \frac{1}{2}\left[u_{i-\frac{1}{2},j} + u_{i+\frac{1}{2},j}\right] \tag{5.77a}$$

$$u_{i+\frac{1}{2},j+\frac{1}{2}} = \frac{1}{2\delta z_{j+\frac{1}{2}}}\left[\delta z_{j+1} u_{i+\frac{1}{2},j} + \delta z_j u_{i+\frac{1}{2},j+1}\right] \tag{5.77b}$$

$$u_{i,j+\frac{1}{2}} = \frac{1}{4\delta z_{j+\frac{1}{2}}}\left[\delta z_{j+1}\left(u_{i-\frac{1}{2},j} + u_{i+\frac{1}{2},j}\right) + \delta z_j\left(u_{i-\frac{1}{2},j+1} + u_{i+\frac{1}{2},j+1}\right)\right] \tag{5.77c}$$

The boundary centered quantities in $y$ direction is $v$ which is defined at $i + \frac{1}{2}, j$. The averaging is as follows.

$$v_{i,j} = \frac{1}{2}\left[v_{i,j-\frac{1}{2}} + v_{i,j+\frac{1}{2}}\right] \tag{5.78a}$$

$$v_{i+\frac{1}{2},j+\frac{1}{2}} = \frac{1}{2\delta r_{i+\frac{1}{2}}}\left[\delta r_{i+1} v_{i,j+\frac{1}{2}} + \delta r_i v_{i+1,j+\frac{1}{2}}\right] \tag{5.78b}$$

$$v_{i+\frac{1}{2},j} = \frac{1}{4\delta r_{i+\frac{1}{2}}}\left[\delta r_{i+1}\left(v_{i,j-\frac{1}{2}} + v_{i,j+\frac{1}{2}}\right) + \delta r_i\left(v_{i+1,j-\frac{1}{2}} + v_{i+1,j+\frac{1}{2}}\right)\right] \tag{5.78c}$$

For continuity equations, the continuity equations is differenced fully implicitly as follows ($k = (f, 1, \ldots, N)$),

$$(\alpha_k\rho_k)_{i,j}^{n+1} = (\alpha_k\rho_k)_{i,j}^{n} - \frac{\delta t}{\delta r_i}[(\alpha_k\rho_k)u_k]_{i,j}^{n+1} - \frac{\delta t}{\delta z_j}[(\alpha_k\rho_k)v_k]_{i,j}^{n+1} \tag{5.79}$$

For momentum equations, the momentum equations are differenced over a staggered mesh using a scheme in which the convection terms are treated explicitly and all other terms are treated implicitly. The difference equations are,

$$
\begin{aligned}
(\alpha_k \rho_k u_k)^{n+1}_{i+\frac{1}{2},j} &= (\overline{\alpha_k \rho_k u_k})_{i+\frac{1}{2},j} - \frac{\delta t}{\delta r_{i+\frac{1}{2}}}\left((p_k)^{n+1}_{i+1,j} - (p_k)^{n+1}_{i,j}\right) \\
&\quad - (w_k)^{n+1}_{i+\frac{1}{2},j} g_r \delta t + \delta t \sum_{\substack{l=f,1 \\ l \neq k}}^{N} (\beta_{lk})^{n}_{i+\frac{1}{2},j}\left((u_l)^{n+1}_{i+\frac{1}{2},j} - (u_k)^{n+1}_{i+\frac{1}{2},j}\right) \\
&\quad + \frac{\delta t}{\delta r_{i+\frac{1}{2}}}\left((\tau_{ck})^{n+1}_{i+1,j} - (\tau_{ck})^{n+1}_{i,j}\right)
\end{aligned} \tag{5.80a}
$$

$$
\begin{aligned}
(\alpha_k \rho_k v_k)^{n+1}_{i,j+\frac{1}{2}} &= (\overline{\alpha_k \rho_k v_k})_{i,j+\frac{1}{2}} - \frac{\delta t}{\delta z_{j+\frac{1}{2}}}\left((p_k)^{n+1}_{i+1,j} - (p_k)^{n+1}_{i,j}\right) \\
&\quad - (w_k)^{n+1}_{i,j+\frac{1}{2}} g_z \delta t + \delta t \sum_{\substack{l=f,1 \\ l \neq k}}^{N} (\beta_{lk})^{n}_{i,j+\frac{1}{2}}\left((v_l)^{n+1}_{i,j+\frac{1}{2}} - (v_k)^{n+1}_{i,j+\frac{1}{2}}\right) \\
&\quad + \frac{\delta t}{\delta z_{j+\frac{1}{2}}}\left((\tau_{ck})^{n+1}_{i,j+1} - (\tau_{ck})^{n+1}_{i,j}\right)
\end{aligned} \tag{5.80b}
$$

where for fluid phase $w_f = \rho_f$ and $\tau_{cf} = 0$, and for particulate phases ($k = 1,\ldots,N$),

$$
w_k = \frac{\alpha_k}{\alpha_f}\left(\rho_k - \sum_{l=f,1}^{N} \alpha_l \rho_l\right) \tag{5.81}
$$

All the explicit terms are lumped as shown below,

$$
\begin{aligned}
(\overline{\alpha_k \rho_k u_k})_{i+\frac{1}{2},j} &= (\alpha_k \rho_k u_k)^{n}_{i+\frac{1}{2},j} - \frac{\delta t}{\delta r_{i+\frac{1}{2}}}[(\alpha_k \rho_k u_k) u_k]^{n}_{i+\frac{1}{2},j} - \frac{\delta t}{\delta z_j}[(\alpha_k \rho_k u_k) v_k]^{n}_{i+\frac{1}{2},j} \\
&\quad + \frac{\delta t}{\delta r_{i+\frac{1}{2}}}\left[(\tau_{krr})^{n}_{i+1} - (\tau_{krr})^{n}_{i,j}\right] + \frac{\delta t}{\delta z_j}\left[(\tau_{krz})^{n}_{i+\frac{1}{2},j+\frac{1}{2}} - (\tau_{krz})^{n}_{i+\frac{1}{2},j-\frac{1}{2}}\right]
\end{aligned} \tag{5.82a}
$$

$$
\begin{aligned}
(\overline{\alpha_k \rho_k v_k})_{i,j+\frac{1}{2}} &= (\alpha_k \rho_k v_k)^{n}_{i,j+\frac{1}{2}} - \frac{\delta t}{\delta r_i}[(\alpha_k \rho_k v_k) u_k]^{n}_{i,j+\frac{1}{2}} - \frac{\delta t}{\delta z_{j+\frac{1}{2}}}[(\alpha_k \rho_k v_k) v_k]^{n}_{i,j+\frac{1}{2}} \\
&\quad + \frac{\delta t}{\delta r_i}\left[(\tau_{krz})^{n}_{i+\frac{1}{2},j+\frac{1}{2}} - (\tau_{krz})^{n}_{i-\frac{1}{2},j+\frac{1}{2}}\right] + \frac{\delta t}{\delta z_{j+\frac{1}{2}}}\left[(\tau_{kzz})^{n}_{i,j+1} - (\tau_{kzz})^{n}_{i,j}\right]
\end{aligned} \tag{5.82b}
$$

As mentioned before, the flux quantities denoted by $\langle \Psi u_k \rangle$ and $\langle \Psi v_k \rangle$ are calculated using donor-cell differencing, where $\Psi$ refers to $(\alpha_k \rho_k)$, $(\alpha_k \rho_k u_k)$ or $(\alpha_k \rho_k v_k)$ quantities. The angular brackets represent donor cell differenced quantities as shown below,

$$\langle \Psi u_k \rangle_{m,p} = (u_k)_{m+\frac{1}{2},p} \begin{cases} (\Psi)_{m,p} & if\,(u_k)_{m+\frac{1}{2},p} \geq 0 \\ (\Psi)_{m+1,p} & if\,(u_k)_{m+\frac{1}{2},p} < 0 \end{cases}$$
$$- (u_k)_{m-\frac{1}{2},p} \begin{cases} (\Psi)_{m-1,p} & if\,(u_k)_{m-\frac{1}{2},p} \geq 0 \\ (\Psi)_{m,p} & if\,(u_k)_{m-\frac{1}{2},p} < 0 \end{cases} \tag{5.83a}$$

$$\langle \Psi v_k \rangle_{m,p} = (v_k)_{m,p+\frac{1}{2}} \begin{cases} (\Psi)_{m,p} & if\,(v_k)_{m,p+\frac{1}{2}} \geq 0 \\ (\Psi)_{m,p+1} & if\,(v_k)_{m,p+\frac{1}{2}} < 0 \end{cases}$$
$$- (v_k)_{m,p-\frac{1}{2}} \begin{cases} (\Psi)_{m,p-1} & if\,(v_k)_{m,p-\frac{1}{2}} \geq 0 \\ (\Psi)_{m,p} & if\,(v_k)_{m,p-\frac{1}{2}} < 0 \end{cases} \tag{5.83b}$$

The viscous stress components are calculated with standard differencing, i.e.,

$$(\nabla \cdot v_k)_{i,j} = \frac{(u_k)_{i+\frac{1}{2},j} - (u_k)_{i-\frac{1}{2},j}}{\delta r_i} + \frac{(v_k)_{i,j+\frac{1}{2}} - (v_k)_{i,j-\frac{1}{2}}}{\delta z_j} \tag{5.84a}$$

$$(\tau_{krr})_{i,j} = 2(\mu_k)_{i,j}\left(\frac{(u_k)_{i+\frac{1}{2},j} - (u_k)_{i-\frac{1}{2},j}}{\delta r_i}\right) + \left(\xi_k - \frac{2}{3}\mu_k\right)_{i,j} (\nabla \cdot v_k)_{i,j} \tag{5.84b}$$

$$(\tau_{kzz})_{i,j} = 2(\mu_k)_{i,j}\left(\frac{(v_k)_{i,j+\frac{1}{2}} - (v_k)_{i,j-\frac{1}{2}}}{\delta z_j}\right) + \left(\xi_k - \frac{2}{3}\mu_k\right)_{i,j} (\nabla \cdot v_k)_{i,j} \tag{5.84c}$$

$$(\tau_{krz})_{i,j} = 2(\mu_k)_{i,j}\left(\frac{(u_k)_{i,j+1} - (u_k)_{i,j-1}}{\delta z_{j-\frac{1}{2}} + \delta z_{j+\frac{1}{2}}} + \frac{(v_k)_{i+1,j} - (v_k)_{i-1,j}}{\delta r_{i-\frac{1}{2}} + \delta r_{i+\frac{1}{2}}}\right) \tag{5.84d}$$

$$(\tau_{krz})_{i+\frac{1}{2},j+\frac{1}{2}} = 2(\mu_k)_{i+\frac{1}{2},j+\frac{1}{2}}\left(\frac{(u_k)_{i+\frac{1}{2},j+1} - (u_k)_{i+\frac{1}{2},j}}{\delta z_{j+\frac{1}{2}}} + \frac{(v_k)_{i+1,j+\frac{1}{2}} - (v_k)_{i,j+\frac{1}{2}}}{\delta r_{i+\frac{1}{2}}}\right) \tag{5.84e}$$

For granular temperature equation, the difference equationusing the donor cell differencing is ($k = 1, 2, \ldots, N$)

$$\frac{2}{3}(\alpha_k \rho_k \theta_k)^{n+1}_{i,j} = \frac{2}{3}(\alpha_k \rho_k \theta_k)^{n}_{i,j} + \left(\overline{\alpha_k \rho_k \theta_k}\right)_{i,j} - \delta t (p_k)^{n+1}_{i,j} (\nabla \cdot v_k)^{n+1}_{i.j}$$
$$+ \delta t (\Phi_k)^{n+1}_{i,j} - \delta t (\gamma_k)^{n}_{i.j} - 3\delta t \left(\beta_{gk}\right)^{n}_{i,j} (\theta_k)^{n+1}_{i.j} \tag{5.85a}$$

$$\left(\overline{\alpha_k \rho_k \theta_k}\right)_{i,j} = \frac{2}{3}\frac{\delta t}{\delta r_i}\left[(\alpha_k \rho_k)^{n+1} (\theta_k)^n (u_k)^{n+1}\right]_{i.j} - \frac{2}{3}\frac{\delta t}{\delta z_j}\left[(\alpha_k \rho_k)^{n+1} (\theta_k)^n (v_k)^{n+1}\right]_{i,j}$$
$$+ \frac{\delta t}{\delta r_i}\left[(k_n)^n_{i+\frac{1}{2},j} \frac{(\theta_k)_{i+1.j} - (\theta_k)_{i.j}}{\delta r_{1+\frac{1}{2}}} - (k_n)^n_{i-\frac{1}{2},j} \frac{(\theta_k)_{i.j} - (\theta_k)_{i-1.j}}{\delta r_{i-\frac{1}{2}}}\right]$$

$$+\frac{\delta t}{\delta x_i}\left[(k_n)^n_{i,j+\frac{1}{2}}\frac{(\theta_k)_{i.j+1}-(\theta_k)_{i.j}}{\delta x_{1+\frac{1}{2}}}-(k_n)^n_{i,j-\frac{1}{2}}\frac{(\theta_k)_{i.j}-(\theta_k)_{i.j-1}}{\delta x_{i-\frac{1}{2}}}\right] \tag{5.85b}$$

For energy balance for gas phase and solids phase, the difference equations using the donor cell differencing can be written as ($k = 1, 2, \ldots, N$)

$$\begin{aligned}(\alpha_g\rho_g H_g)^{n+1}_{i,j} &= (\alpha_g\rho_g H_g)^n_{i,j} + \left(\widetilde{\alpha_g\rho_g H_g}\right)_{i,j} + \delta t(h_{gk})^n_{i,j}(T_k - T_g)^{n+1}_{i.j}\\ &+ \delta t(\beta_{gk})^n_{i,j}\{((u_k)^n_{i.j} - (u_g)^n_{i.j})^2 + ((v_k)^n_{i.j} - (v_g)^n_{i.j})^2\} + \delta t(\dot{Q}_{re,g})^n_{i,j}\end{aligned} \tag{5.86a}$$

$$(\alpha_k\rho_k H_k)^{n+1}_{i,j} = (\alpha_k\rho_k H_k)^n_{i,j} + \left(\widetilde{\alpha_k\rho_k H_k}\right)_{i,j} + \delta t(h_{gk})^n_{i,j}(T_g - T_k)^{n+1}_{i.j} + \delta t(\dot{Q}_{re,k})^n_{i,j} \tag{5.86b}$$

The "tilde" quantities for gas phase and solids phaseare given by,

$$\begin{aligned}\left(\widetilde{\alpha_g\rho_g H_g}\right)_{i,j} &= \frac{\delta t}{\delta r_i}\left[(\alpha_g\rho_g)^{n+1}(H_g)^n(u_g)^{n+1}\right]_{i.j} - \frac{\delta t}{\delta z_j}\left[(\alpha_g\rho_g)^{n+1}(H_g)^n(v_g)^{n+1}\right]_{i,j}\\ &+ ((p_g)^{n+1}_{i,j} - (p_g)^n_{i,j}) + \frac{\delta t}{\delta r_i}(u_g)^{n+1}_{i,j}\left[(p_g)^{n+1}_{i+\frac{1}{2},j} - (p_g)^{n+1}_{i-\frac{1}{2},j}\right]\\ &- \frac{\delta t}{\delta x_i}(v_g)^{n+1}_{i,j}\left[(p_g)^{n+1}_{i,j+\frac{1}{2}} - (p_g)^{n+1}_{i,j-\frac{1}{2}}\right]\end{aligned} \tag{5.87a}$$

$$\begin{aligned}\left(\widetilde{\alpha_k\rho_k H_k}\right)_{i,j} &= \frac{\delta t}{\delta r_i}\left[(\alpha_k\rho_k)^{n+1}(H_k)^n(u_k)^{n+1}\right]_{i.j} - \frac{\delta t}{\delta z_j}\left[(\alpha_k\rho_k)^{n+1}(H_k)^n(v_k)^{n+1}\right]_{i,j}\\ &+ \frac{\delta t}{\delta r_i}\left[(k_k)^n_{i+\frac{1}{2},j}\frac{(T_k)_{i+1,j}-(T_k)_{i,j}}{\delta r_{i+\frac{1}{2}}} - (k_k)^n_{i-\frac{1}{2},j}\frac{(T_k)_{i,j}-(T_k)_{i-1,j}}{\delta r_{i-\frac{1}{2}}}\right]\\ &+ \frac{\delta t}{\delta x_i}\left[(k_k)^n_{i,j+\frac{1}{2}}\frac{(T_k)_{i,j+1}-(T_k)_{i,j}}{\delta x_{i+\frac{1}{2}}} - (k_k)^n_{i,j-\frac{1}{2}}\frac{(T_k)_{i,j}-(T_k)_{i,j-1}}{\delta x_{i-\frac{1}{2}}}\right]\\ &+ \delta t(\Phi_k)^n_{i,j}\end{aligned} \tag{5.87b}$$

From the species balance for gas and solids components ($k = 1, 2, \ldots, N$), the difference forms of species equation can be written as

$$\begin{aligned}(\alpha_g\rho_g Y_k)^{n+1}_{i,j} &= (\alpha_g\rho_g Y_k)^n_{i,j} - \frac{\delta t}{\delta r_i}[(\alpha_g\rho_g Y_k)u_g]^{n+1}_{i,j} - \frac{\delta t}{\delta x_i}[(\alpha_g\rho_g Y_k)v_g]^{n+1}_{i,j}\\ &+ \delta t(\dot{m}^k_g)^n_{i,j}\end{aligned} \tag{5.88a}$$

$$\begin{aligned}(\alpha_k\rho_k X_k)^{n+1}_{i,j} &= (\alpha_k\rho_k X_k)^n_{i,j} - \frac{\delta t}{\delta r_i}[(\alpha_k\rho_k X_k)u_k]^{n+1}_{i,j} - \frac{\delta t}{\delta x_i}[(\alpha_k\rho_k X_k)v_k]^{n+1}_{i,j}\\ &+ \delta t(\dot{m}_k)^n_{i,j}\end{aligned} \tag{5.88b}$$

The finite difference forms of balance equations form the coupled non-linear algebraic relations for cell variables of gas pressure $p$, gas density $\rho_g$, gas volume fraction $\alpha_g$, granular temperature $\theta$, and temperature $T$ and velocity of gas phase $u_g$ and solids phase $u_k$. The finite difference forms of species balance equation are solved simultaneously to obtain weight fractions of each gas species $Y_k$ and solids component $X_k$. The calculations start with a known initial distribution of the above variables. At each computational cycle their values are updated by solving the finite difference equations.

### 5.5.5 Solution Technique for Finite-Difference Equations

For solution of the momentum equations, the equations are recast in the following formto facilitate the particular method of solution. The momentum equation in –$r$ direction could be collected together in a matrix form (Ettehadieh 1982; Syamlal 1985; Gamwo 1992).

$$(A)_{i+\frac{1}{2},j}(U)^n_{i+\frac{1}{2},j} = (B_u)_{i+\frac{1}{2},j} \tag{5.89a}$$

$$A = \begin{pmatrix} A_{ff} & A_{f1} & A_{f3} & \cdots & A_{fN} \\ A_{1f} & A_{11} & A_{12} & \cdots & A_{1N} \\ \vdots & \vdots & \ddots & \ddots & \vdots \\ A_{Nf} & A_{N1} & A_{N2} & \cdots & A_{NN} \end{pmatrix} \tag{5.89b}$$

where

$$A_{kk} = (\varepsilon_k \rho_k)^{n+1} + \delta t \sum_{\substack{l=f,l \\ l \neq k}}^{N} (\beta_{lk})^n \tag{5.90a}$$

$$A_{kl} = A_{lk} = -\delta t(\beta_{lk})^n \tag{5.90b}$$

$$A_{i+\frac{1}{2},j} \cdot \begin{pmatrix} u_f \\ u_1 \\ \vdots \\ u_N \end{pmatrix} = \begin{pmatrix} (\overline{\alpha_f \rho_f u_f}) - \frac{\delta t}{\delta r_{i+\frac{1}{2}}}\left((p_f)^{n+1}_{i+1,j} - (p_f)^{n+1}_{i,j}\right) - (\alpha_f \rho_f) g_r \delta t \\ (\overline{\alpha_1 \rho_1 u_1}) - \frac{\delta t}{\delta r_{i+\frac{1}{2}}}\left((p_1)^{n+1}_{i+1,j} - (p_1)^{n+1}_{i,j}\right) - (\alpha_1 \rho_1) g_r \delta t \\ \vdots \\ (\overline{\alpha_N \rho_N u_N}) - \frac{\delta t}{\delta r_{i+\frac{1}{2}}}\left((p_N)^{n+1}_{i+1,j} - (p_N)^{n+1}_{i,j}\right) - (\alpha_N \rho_N) g_r \delta t \end{pmatrix} \tag{5.90c}$$

are similarly, momentum equation in $z$-direction can be written as,

$$(A)_{i,j+\frac{1}{2}}(V)^{n}_{i,j+\frac{1}{2}} = (B_v)_{i,j+\frac{1}{2}} \tag{5.91a}$$

$$A_{i,j+\frac{1}{2}} \cdot \begin{pmatrix} v_f \\ v_1 \\ \vdots \\ v_N \end{pmatrix} = \begin{pmatrix} \left(\overline{\alpha_f \rho_f v_f}\right) - \frac{\delta t}{\delta z_{j+\frac{1}{2}}}\left((p_f)^{n+1}_{i,j+1} - (p_f)^{n+1}_{i,j}\right) - (\alpha_f \rho_f) g_z \delta t \\ \left(\overline{\alpha_1 \rho_1 v_1}\right) - \frac{\delta t}{\delta z_{j+\frac{1}{2}}}\left((p_1)^{n+1}_{i,j+1} - (p_1)^{n+1}_{i,j}\right) - (\alpha_1 \rho_1) g_z \delta t \\ \vdots \\ \left(\overline{\alpha_N \rho_N v_N}\right) - \frac{\delta t}{\delta z_{j+\frac{1}{2}}}\left((p_N)^{n+1}_{i,j+1} - (p_N)^{n+1}_{i,j}\right) - (\alpha_N \rho_N) g_z \delta t \end{pmatrix} \tag{5.91b}$$

For the convergence on fluid continuity equation, the solution process is carried out in two major steps. First of all, the continuity equations, the momentum equations, and a part of energy equation are solved simultaneously to establish the pressure and the velocity fields. In this step only the interphase heat transfer part of the energy equation is considered. Secondly, the remaining parts of the energy equations are solved to establish the temperature profiles. The solution procedure of computation sweep is illustrated in Fig. 5.7 (Sun 1996; Wu 1996; Diana 2000; Mostofi 2002; Kashyap 2010).

The first step proceeds as follows:

$$\alpha_f = 1 - \sum_{k=1}^{N} \alpha_k \tag{5.92a}$$

$$D_{i,j} = -\left(\alpha_f \rho_f\right)^{n+1}_{i,j} + \left(\alpha_f \rho_f\right)^{n}_{i,j} - \frac{\delta t}{R_i \delta r_i}\left[R\left(\alpha_f \rho_f\right) u_f\right]^{n+1}_{i,j} - \frac{\delta t}{\delta z_j}\left[\left(\alpha_f \rho_f\right) v_f\right]^{n+1}_{i,j} \tag{5.92b}$$

$$D_{i,j} \le CONV^{n+1}_{i,j} = EPSG\left(\alpha_f \rho_f\right)^{n}_{i,j} \tag{5.92c}$$

For pressure iteration, the pressure is adjusted using a combination of Newton's method and secant method when $D_{i,j}$ fails to meet the convergence criterion in any cell. The initial adjustment of pressure uses Newton's method.

$$\left(p_f\right)^{m+1} = \left(p_f\right)^{m} - \omega \frac{D^m}{\partial D / \partial \left(p_f\right)^{m}} \tag{5.93}$$

where the indices $i$, $j$ and $n$ have been omitted. The index, $m$, indicates the iteration level. This is equivalent to using Newton's method for each cell, where $\omega$ is a relaxation parameter near unity, and $\overline{\beta}$ is computed as,

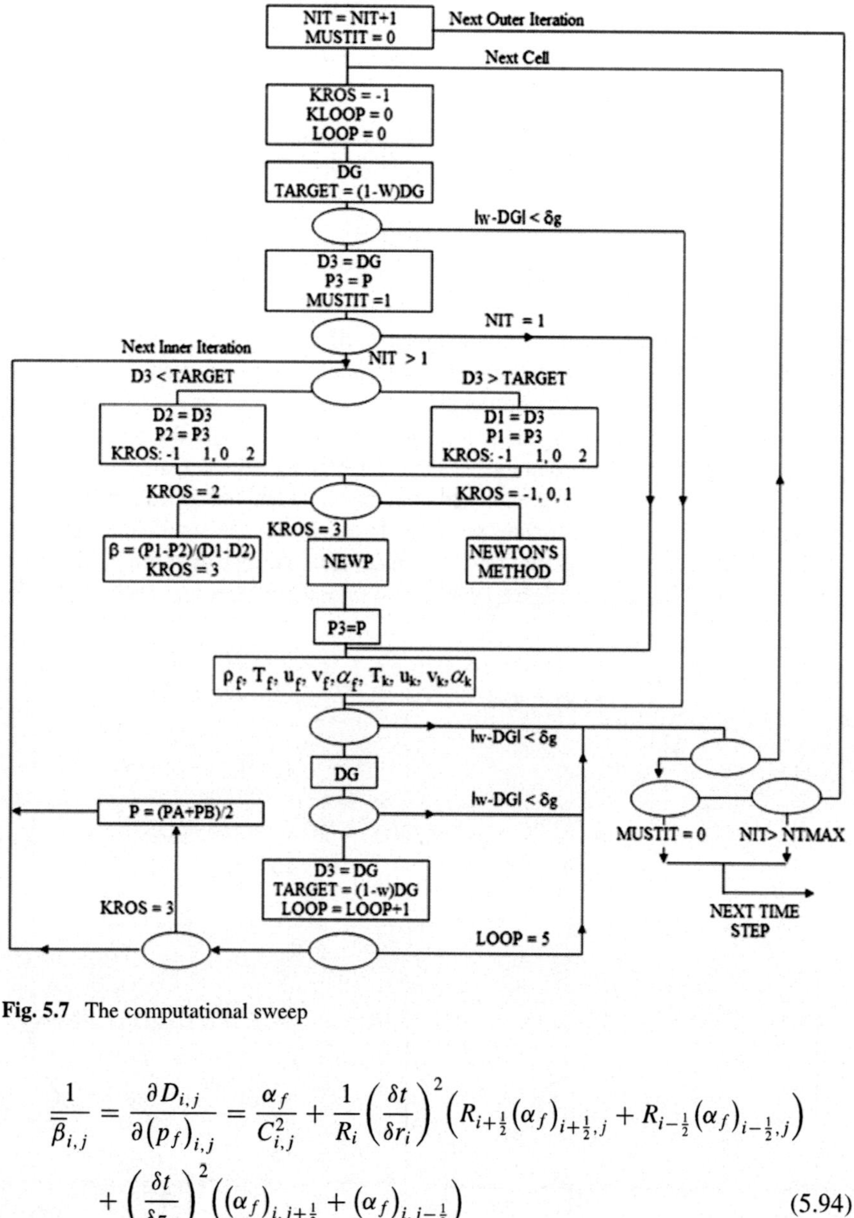


**Fig. 5.7** The computational sweep

$$\frac{1}{\overline{\beta}_{i,j}} = \frac{\partial D_{i,j}}{\partial (p_f)_{i,j}} = \frac{\alpha_f}{C_{i,j}^2} + \frac{1}{R_i}\left(\frac{\delta t}{\delta r_i}\right)^2 \left(R_{i+\frac{1}{2}}(\alpha_f)_{i+\frac{1}{2},j} + R_{i-\frac{1}{2}}(\alpha_f)_{i-\frac{1}{2},j}\right)$$
$$+ \left(\frac{\delta t}{\delta z_j}\right)^2 \left((\alpha_f)_{i,j+\frac{1}{2}} + (\alpha_f)_{i,j-\frac{1}{2}}\right) \tag{5.94}$$

once every time step.

The sound speed $C_{i,j}$ is given by,

$$C_{i,j}^2 = \left(\frac{\partial p_f}{\partial \rho_f}\right)_{i,j} \tag{5.95}$$

where $\left(\partial p_f/\partial \rho_f\right)$ can be determined from the equation of state. This formation is only approximate. Hence, subsequent use the secant method:

$$\left(p_f\right)^{m+1} = \left(p_f\right)^m - \omega\left(\frac{\left(p_f\right)^{m-1} - \left(p_f\right)^m}{D^{m-1} - D^m}\right)D^m \tag{5.96}$$

The use of secant method is combined until $D_{i,j}$ changes sign. Thereafter a combination of the secant method and a bisection method is used. The method is illustrated in the Fig. 5.8. Given the three pressure $p_1$, $p_2$ and $p_3$ of which $p_1$ and $p_2$ bracket the desired pressure and $p_3$ lies between them and the respective mass residuals $D_1$, $D_2$ and $D_3$ do not satisfy the convergence criterion in cell $(i, j)$, $D_1 > 0$, and $D_2 < 0$. With three pressures and their mass residuals obtained as described, or otherwise a constrained two-sides secant technique is used to obtain further pressure adjustment. Form these pressures and their mass residuals, the pressure $p_A$ and $p_B$ are determined by straight line extrapolation and interpolation, respectively, as follow,

$$p_A = \begin{cases} (p_3D_1 - p_1D_3)/(D_1 - D_3) & \text{for} \quad D_1 \neq D_3 \\ (p_2 + p_3)/2 & \text{for} \quad D_1 = D_3 \end{cases} \tag{5.97}$$

and

$$p_B = \begin{cases} (p_3D_2 - p_2D_3)/(D_2 - D_3) & \text{for} \quad D_2 \neq D_3 \\ (p_1 + p_3)/2 & \text{for} \quad D_2 = D_3 \end{cases} \tag{5.98}$$

The new estimate of the advanced time pressure is then computed as,

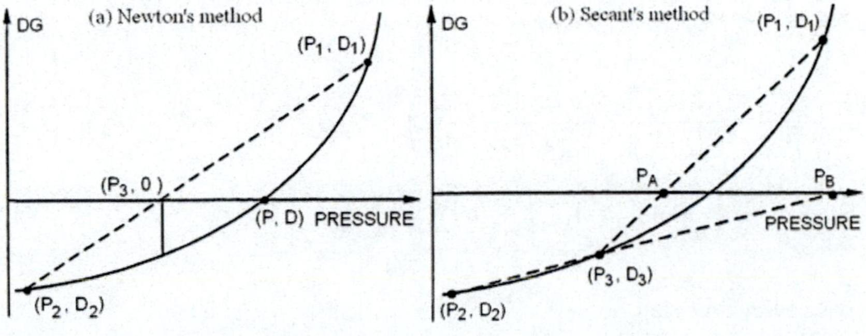


Fig. 5.8 The methods for pressure iteration

$$\left(p_f\right)^{m+1} = \frac{1}{2}(p_A + p_B) \tag{5.99}$$

If the pressure, $p_A$ should lie outside the interval $p_1$ to $p_3$, it is given the valve $\frac{1}{2}(p_A + p_B)$. After $\left(p_f\right)^{m+1}$ is estimated, point 2 is discarded add points 1 and 3 are retained as improved bounds for the next pressure estimate. When $D_{i,\,j}$ changes sign, the valve of $\overline{\beta}$ is also updated for future iteration as,

$$\overline{\beta} = \frac{p_1 - p_2}{D_1 - D_2} \tag{5.100}$$

## 5.6 Boundary Conditions

### *5.6.1 Wall Boundary Conditions*

The rectangular region in, which calculations are to preformed is partitioned into cells of sizes $dx_i$ (or $dr_i$) in $x$ (or $r$) -direction and $dy_i$ (or $dz_i$) in $y$ (or $z$) -direction. A perimeter of fictitious (dummy) boundary cells surrounding the computing mesh is used to enforce boundary conditions. Several boundary conditions around the computing mesh perimeter are programmed in the code. The free and no slip boundary conditions are shown in Fig. 5.9.

Three types of boundary conditions may be specified for a rigid cell: free-slip, no-slip (marked by IFL $= 2$) and partial slip (marked by IFL $=$ 3). In two dimensions, a free-slip boundary represents line of symmetry and a non-adhering boundary that exerts no drag on the fluid; a no-slip boundary represents a viscous boundary that exerts a drag on the fluid.

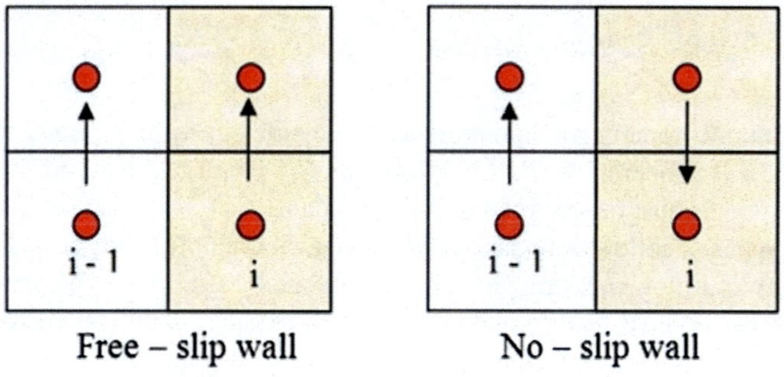


**Fig. 5.9** Free and no slip conditions

Consider cell $(i,\ j)$, which is a fluid cell, an inflow boundary cell, or an outflow boundary cell. An adiabatic rigid cell is said to be a corner cell if and only if it has at least two adjacent edged, each of which is shared with afluid cell, an inflow boundary cell, or an outflow boundary cell.

If right cell $(i\ +\ 1,\ j)$ or left cell $(i\ -\ 1,\ j)$, is a rigid cell, then for all time levels $n$,

$$(u_k)^{n+1}_{i\pm\frac{1}{2},j} = 0 \tag{5.101a}$$

$$(v_k)^{n+1}_{i\pm1,j+\frac{1}{2}} = \begin{cases} (v_k)^n_{i,j+\frac{1}{2}} & \text{for } IFL = 2 \\ -\delta r_{i\pm1}(v_k)^n_{i,j+\frac{1}{2}}/\delta r_i & \text{for } IFL = 3, k \neq \text{solid} \\ \left(\frac{2\lambda_{i,j}-\delta r_{i\pm1}}{2\lambda_{i,j}-\delta r_i}\right)(v_k)_{i,j+\frac{1}{2}} & \text{for } IFL = 3, k = \text{solid} \end{cases} \tag{5.101b}$$

Similarly, if top cell $(i,\ j\ +\ 1)$ or bottom cell $(i,\ j\ -\ 1)$, is a rigid cell and if free-slip boundary conditions (IFL $=$ 2) or no-slip boundary conditions (IFL $=$ 3) are imposed, then for all $n$,

$$(u_k)^{n+1}_{i+\frac{1}{2},j\pm1} = \begin{cases} (u_k)^n_{i+\frac{1}{2},j} & \text{for } IFL = 2 \\ -\delta z_{j\pm1}(u_k)^n_{i+\frac{1}{2},j}/\delta z_j & \text{for } IFL = 3, k \neq \text{solid} \\ \left(\frac{2\lambda_{i,j}-\delta z_{j\pm1}}{2\lambda_{i,j}-\delta z_j}\right)(u_k)_{i+\frac{1}{2},j} & \text{for } IFL = 3, k = \text{solid} \end{cases} \tag{5.102a}$$

$$(v_k)^{n+1}_{i,\ j\pm\frac{1}{2}} = 0 \tag{5.102b}$$

If any one of the cells $(i\ +\ 1,\ j)$, $(i\ -\ 1,\ j)$, $(i,\ j\ +\ 1)$, or $(i,\ j\ -\ 1)$ is both a rigid cell and a corner cell, then, for all free-slip, no-slip and particle-slip boundary condition, all its velocity components located at the center are set equal to zero.

### 5.6.2 Inflow Boundary Cells

For each inflow opening a fluid pressure, $p_f$, velocities $u_k$ and $v_k$, and solid volume fraction, $\alpha_k$, temperature, $T_k$ for all phases ($k = f$, 1,..., $N$), and solid granular temperature $\theta$ must be specified in the input data.

The types of inflow boundary conditions used are inflow prescribed (marked by IFL $=$ 5); or inflow pressure prescribed, (marked by IFL $=$ 6). The pressure, $p_f$, is required for both (IFL $=$ 5) and (IFL $=$ 6) to compare the mass fluxes. However, when the inflow pressure is prescribed, the radial or axial velocity components are computed using momentum equations.

### 5.6.3 Outflow Boundary Cells

For each outflow opening, a pressure, $p_f$, is specified at the beginning of each computation cycle. The types of outflow boundary conditions used are: pressure prescribed outflow with particulates outflow (marked by IFL = 7), or no particulates outflow (marked by IFL = 8), or continuative outflow (marked by IFL = 4).

For outflow boundary conditions, the volume fractions, for a given cell of an outflow opening, is obtaining by reflection. The tangential components of the velocities are set to zero. A numerical screen is used to keep particulate phases form leaving the fluidized bed for (IFL = 8). Therefore, at the top outflow boundary, where $(i,\ j)$ = (I, JB2); $\left(i+\frac{1}{2},\ j\right)$ = (I, JB2); $\left(i, j+\frac{1}{2}\right)$ = (I, JB2); $(i,\ j-1)$ = (I, JB1) and $\left(i,\ j-\frac{1}{2}\right)$ = (I, JB1), we have,

$$\left(p_f\right)_{i,j}^{n+1}=\begin{cases} p_o & \text{for } IFL=7 \text{ or } 8 \\ \left(p_f\right)_{i,j-1}^{n} & \text{for } IFL=4 \end{cases} \tag{5.103a}$$

$$(\alpha_k)_{i,j}^{n+1}=\begin{cases} (\alpha_k)_{i,j-1}^{n} & \text{for } IFL=4 \text{ or } 7 \\ \begin{cases} 1 & k=f \\ 0 & k\neq f \end{cases} & \text{for } IFL=8 \end{cases} \tag{5.103b}$$

$$(u_k)_{i+\frac{1}{2},j}^{n+1}=0 \tag{5.103c}$$

$$\left(v_f\right)_{i,j+\frac{1}{2}}^{n+1}=\left(\alpha_f\rho_f\right)_{i,j-1}^{n+1}\left(v_f\right)_{i,j-\frac{1}{2}}^{n+1}/\left(\alpha_f\rho_f\right)_{i,j}^{n+1} \tag{5.103d}$$

$$(v_k)_{i,j+\frac{1}{2}}^{n+1}=\begin{cases} (\alpha_k\rho_k)_{i,j-1}^{n+1}(v_k)_{i,j-\frac{1}{2}}^{n+1}/(\alpha_k\rho_k)_{i,j}^{n+1} & \text{for } IFL=4 \text{ or } 7 \\ 0 & \text{for } IFL=8 \end{cases} \tag{5.103e}$$

At the right outflow boundary, where $(i,\ j)$ = (IB2, J); $\left(i+\frac{1}{2},\ j\right)$ = (IB2, J); $(i+1,\ j)$ = (IB1, J), $\left(i-\frac{1}{2},\ j\right)$ = (IB1, J), $(i-1,\ j)$ = (IB1, J), and we have,

$$\left(p_f\right)_{i,j}^{n+1}=\begin{cases} p_o & \text{for } IFL=7 \text{ or } 8 \\ \left(p_f\right)_{i-1,j}^{n} & \text{for } IFL=4 \end{cases} \tag{5.104a}$$

$$(\alpha_k)_{i,j}^{n+1}=\begin{cases} (\alpha_k)_{i-1,j}^{n} & \text{for } IFL=4 \text{ or } 7 \\ \begin{cases} 1 & k=f \\ 0 & k\neq f \end{cases} & \text{for } IFL=8 \end{cases} \tag{5.104b}$$

$$\left(u_f\right)_{i+\frac{1}{2},j}^{n+1}=r_{i-\frac{1}{2}}\left(\alpha_f\rho_f\right)_{i-1,j}^{n+1}\left(u_f\right)_{i-\frac{1}{2},j}^{n+1}/\left[r_{i+\frac{1}{2}}\left(\alpha_f\rho_f\right)_{i,j}^{n+1}\right] \tag{5.104c}$$

$$
(u_k)^{n+1}_{i+\frac{1}{2},j} = \begin{cases} r_{i-\frac{1}{2}}(\alpha_k\rho_k)^{n+1}_{i-1,j}(u_k)^{n+1}_{i-\frac{1}{2},j}/\left[r_{i+\frac{1}{2}}(\alpha_k\rho_k)^{n+1}_{i,j}\right] & \text{for } IFL = 4 \text{ or } 7 \\ 0 & \text{for } IFL = 8 \end{cases} \tag{5.104d}
$$

## 5.6.4 Initial Conditions

At the beginning of the simulation, the distribution of all field variables is given by the initial conditions. Uniform and simple non-uniform initial conditions can be specified using the input data. Fluid cell (marked by IFL = 1) with the filed variables may be defined in similar manner as the obstacle blocks are defines. Complex non-uniform initial conditions should be programmed.

The pressure in the axial direction (marked by IPRES) is initialized by the gravity head of the fluid or/and particulate phase. If (IPRES = 0), pressure profile is obtained using fluid phase only and the weight of the particulate phases is not supported by fluid phase. If (IPRES = 1), the fluid phase supports the weight of the computed completed bed.

The pressure distribution in all fluid cells is computed recursively from the weight of the bed as follows,

$$
(p_f)_{i,j} = (p_f)_{i+1,j} - g_z\delta z_{j+\frac{1}{2}} \begin{cases} (\alpha_f\rho_f)_{i,j+\frac{1}{2}} & \text{for } IPRES = 0 \\ \sum\limits_{k=f,1}^{N}\left[(\alpha_k\rho_k)_{i,j+\frac{1}{2}}\right] & \text{for } IPRES = 1 \end{cases} \tag{5.105}
$$

where pressure in the most top cell J = JB2 is given by $(p_f)_{i,j+\frac{1}{2}} = p_0$. For density of fluid, equation of state or any equivalent relation may be used. Since this density may depend on the pressure in cell (*i*, *j*) which is yet to be calculated, a quadratic equation must be solved in a general case.

## 5.6.5 Tips for Running the CFD Code

First of all, choose an appropriate coordinate system from rectangular coordinates. Specify $\delta r_i$ and $\delta z_j$ according to the degree of resolution required. Sometimes these dimensions are governed by the sized of inflow and outflow opening and also boundary layers. However, the ratio between two adjacent increments should be within bounds,

$$
\frac{1}{\overline{K}} < \frac{\delta r_i}{\delta r_{i+1}} \quad \text{or} \quad \frac{\delta z_j}{\delta z_{j+1}} < \overline{K} \tag{5.106}
$$

where $\overline{K}$ is usually a number between 1 and 2. The accuracy (and occasionally the stability) is sensitive to the value of $\overline{K}$ used. For satisfactory results $\overline{K} \leq 1.15$ should be used. But for large system, a larger value of $\overline{K}$ around 1.8 is recommended in order to get a reasonable CPU time.

Use the symmetries of the problem to reduce the computational region. Properly specify the inflow and outflow coordinates which can be located in any positions.

The time step $\delta t$ is limited by Courant stability condition (Courant et al. 1952)

$$\delta t < \frac{\min(\delta r_i, \delta z_j)}{\max(velocity)} \tag{5.107}$$

The solution technique is a generalization of the Implicit Continuum Eulerian (ICE) method and hence here is no sonic time step limitation. However, in some cases the time step required for proper convergence could be much less than what is given by the above expression. If the system cannot meet the convergence criterion, one should reduce the time step until the convergence criterion is reached.

Once the output file comes out, check the system geometry given in the output file. If the geometry is not correct, go back to check the input data, there must be something wrong with either input data file.

## References


Anderson, J. D. (1995). *Computational fluid dynamics: The Basics with Applications.* New York: McGraw-Hill.

Aris, R. (1962). *Vectors, tensors, and the basic equations of fluid mechanics.* Englewood Cliffs NJ Prentice-Hall.

Courant, R., & Hilbert, D. (1962). Methods of mathematical physics Vol. II: Partial Differential Equations. New York: Wiley-Interscience.

Courant, R., Iasscson, E., & Rees, M. (1952). On the solution of nonlinear hyperbolic differential equations by finite differences. *Communication in Pure and Applied Mathematics, 5,* 243-255.

Diana, M. (2000). Hydrodynamic simulation of a gas-liquid-solid fluidized bed. Ph.D. Thesis, Illinois Institute of Technology, Chicago.

Ettehadieh, B. (1982). Hydrodynamic analysis of gas-solid fluidized beds. Ph.D. Thesis, Illinois Institute of Technology, Chicago.

Gamwo I (1992) Multi-dimensional hydrodynamic and erosion modeling of fluidized bed. Ph.D. Thesis, Illinois Institute of Technology, Chicago

Gidaspow, D. (1994). *Multiphase flow and fluidization: Continuum and kinetic theory descriptions.* California: Academic Press.

Harlow, F. H., & Amdsen, A. A. (1971). A numerical fluid dynamic calculation method for all flow speeds. *Journal of Computational Physics, 8,* 197–213.

Kashiwa, B. A., Padial, N. T., Rauenzahn, R. M., & Vanderheyden, W. B. (1994). A cell-centered ICE method for multiphase flow simulation. Los Alamos National Laboratory Research Report, LA-UR-93–3922.

Kashyap, M. (2010). Measurements and computation of dispersion and mass transfer coefficient in fluidized beds. Ph.D. Thesis, Illinois Institute of Technology, Chicago.

Mostofi, R. (2002). CFD simulations of particulate two-phase and three-phaseflows. Ph.D. Thesis, Illinois Institute of Technology, Chicago.

Patankar, S. V. (1980). *Numerical heat transfer and fluid flow*. Hemisphere Publishing Corp: Taylor and Francis.

Rivard, W. C., & Torrey, M. D. (1977). K-FIX: A computer program for transient, two dimensional, two-fluid flow. Los Alamos Scientific Laboratory, LA-Nureg-6623:NRC-4.

Sun, B. (1996). Simulation of gas-liquid and gas-solid two-phase flows.Ph.D. Thesis, Illinois Institute of Technology, Chicago, IL.

Syamlal, M. (1985). Multiphase hydrodynamics of gas-solid flow. Ph.D. Thesis, Illinois Institute of Technology, Chicago.

Wu, Y. (1996). Simulation of methanol synthesis in slurry bubble column reactors. Ph.D. Thesis, Illinois Institute of Technology, Chicago.

# Chapter 6
# Cases for Numerical Simulations of Fluidized Bed Systems

## Nomenclature

| | |
|---|---|
| $a$ | Absorption coefficient, 1/m |
| $C$ | Gas concentration, kmol/m$^3$ |
| $c$ | Specific heat capacity, kJ/kg K |
| $D_g$ | Diffusivity coefficient, m$^2$/s |
| $d_p$ | Particle diameter, m |
| $e$ | Emissivity |
| $E$ | Activation energy, J/kg mol |
| $G$ | Incident radiation, W/m$^2$ |
| $G_s$ | Solid flux, kg/m$^2$ s |
| $k$ | Kinetic rate, 1/s |
| $M$ | Molecular weight, kg/mol |
| $q$ | Heat flux, W/m$^2$ |
| $p$ | Pressure, Pa |
| $Pr$ | Prandtl number |
| $R$ | Universal gas constant, kJ/mol K |
| $R_i$ | Reaction rate, mol/m$^3$s |
| $Re$ | Reynolds number |
| $T$ | Temperature, K |
| $u$ | Superficial gas velocity, m/s |
| $X$ | Char mass fraction, kg-char/kg-bed material |
| $Y$ | Mass fraction of gas species, kg-gas species/kg-gas |
| $Y_i$ | Concentration of gas species $i$, kmol/m$^3$ |
| z | Distance, m |

## Greek Symbols

| | |
|---|---|
| $\alpha$ | Volume fraction |
| $\mu_g$ | Gas viscosity, kg/ms |



H. Lu et al., *Computational Fluid Dynamics and the Theory of Fluidization*,
https://doi.org/10.1007/978-981-16-1558-0_6

$\rho$ Density, kg/m$^3$
$\sigma$ Stefan–Boltzmann constant, W/m$^2$ K$^4$

**Subscripts**

$g$ Gas phase
$i$ Gas species i
$s$ Solids phase

Gas-particles two-fluid flow macroscopic modeling (TFM) is currently considered to be the most suitable type of modeling for fluidized beds under conditions that are relevant for industrial systems. The guidelines and examples of how these different sub-models can be linked into a comprehensive process model are given. This chapter is limited to modeling that is specific for fluidized bed processes, i.e., that focuses mainly on the gas-solids side. Thus, general models required for full process simulations are not considered. In general, the goal of modeling is to provide information that can be used for the reliable design, scale-up, and process optimization of fluidized bed systems.

## 6.1 Fluidized Bed Coal Combustors

The energy resources available in nature such as coal and biomass cannot be used directly for most of the applications. These solids fuels have to be transformed into a useful form such as heat or electricity before their use. This indicates that the coal must first be burned or combusted to generate heat which is then used to produce steam for space heating and electricity generation.

Fluidized bed combustors use coal and other solid fuels of biomass and agricultural materials as the primary source for power generation applications since they offer fuel flexibility, and broad turn down ratio and relatively fast response to load changes. The possibility of capturing sulphur oxides emitted during the fluidized bed combustion of solid fuels using calcium-based sorbents (such as limestone) has been widely explored and demonstrated. Interesting alternative application fields for the municipal waste wastes can be re-use in the low-energy industry. Solid fuels react either with oxygen to produce heat (combustion), or with $H_2O$, $CO_2$ or their mixtures to produce a syngas with an appreciable heating value (gasification) in fluidized bed systems. Hence, the understanding of flow behavior of gas-particles mixture and chemical reactions is required for better design of heat transferring surfaces in the fluidized bed combustor.

### 6.1.1 Circulating Fluidized Bed (CFB) Coal Combustors

Figure 6.1 shows a schematic of CFB boiler with steam capacity of 220 t/h. The combustor is 8.61 × 4.77 m in cross-section and 31.46 m high (The height is defined in this study as the distance above the distributor). It is equipped with two hot cyclones for particles separation and two loop-seals for particles recirculation. The cross-section of furnace area increases gradually from 2.5 $m^2$ at the distributor to 41.07 $m^2$ in the furnace with the diverging section ending 3.23 m above the distributor. The flue gases and entrained particles pass from the top of the combustion chamber into cyclones which capture the solid particles and return them to the bottom of the combustion chamber. Two hot cyclones (barrel diameter: 4.0 m, barrel length: 3.3 m, length of cone section: 5.8 m and dip leg diameter: 0.35 m) are located symmetrically at the rear wall at the top of the combustor. Two recycle inlets for return of particles collected by cyclones penetrate the rear wall in the lower part of the combustor. Coal is fed there together with recycled solids.

Initial particles are set up a bed materials zone. The static height of bed materials and the mass of particles in the return leg keep a constant. Table 6.1 summarizes the parameters used in the simulations (Guodong et al. 2010). At the inlet, the velocities of the primary air and particles are specified. The concentration of particles is to be zero. The inlet mass flux of gas is specified at the loop seal. The gas pressure is not specified at the inlet because of the incompressible gas phase assumption (relatively low pressure drop system). At the outlet, the pressure is specified (atmospheric). Initially, the velocities of both the gas and particles are set at zero. At the wall, the gas tangential and normal velocities are set at zero. The normal velocity of particles is also set at zero.

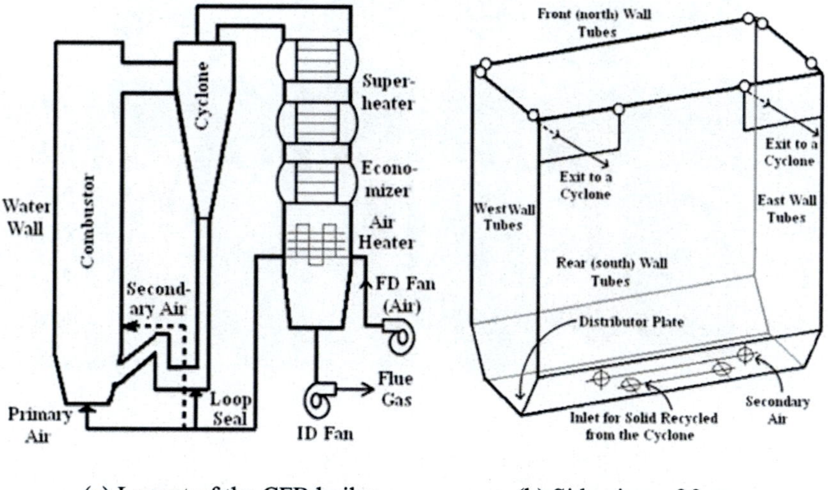


(a) Layout of the CFB boiler (b) Side view of furnace

**Fig. 6.1** Schematic of the circulating fluidized bed combustor

**Table 6.1** Parameters and dimensions of furnace of the model used

| Parameter | Unit | Cases | Parameter | Unit | Cases |
|---|---|---|---|---|---|
| Height of furnace | m | 31.46 | Diameter of particles | mm | 0.5 |
| Width of furnace | m | 4.77 | Density of particles | kg/m$^3$ | 2200 |
| Depth of furnace | m | 8.61 | Operating temperature | °C | 850 |
| Outer diameter of cyclone | m | 4.0 | Gas density | kg/m$^3$ | 0.315 |
| Superficial velocity | m/s | 5.0 | Specularity coefficient | / | 0.5 |
| Particle–particle coefficient of restitution | / | 0.95 | Maximum solid volume fraction | / | 0.62 |
| Wall–particle coefficient of restitution | / | 0.95 | Static height of particles | m | 3.0, 5.0 |

Figure 6.2 shows the instantaneous concentrations of particles in the CFB combustor. Simulations show that the bed begins to expand once gas is introduced. Some particles hit the top wall and fall down. Some particles close to the cyclone are dragged by gas into the cyclone. At the initial stage, the flow of particles into the cyclone is high, leading to particle accumulation at the apex of the cyclone. Some particles may exit the system from vortex finder due to strong particle–particle interaction. At the same time, the equal mass flux of particles will be added into the furnace from loop seal to keep the constant inventory. Finally, macroscopically steady state flow is established, where the input and output rates of particles are

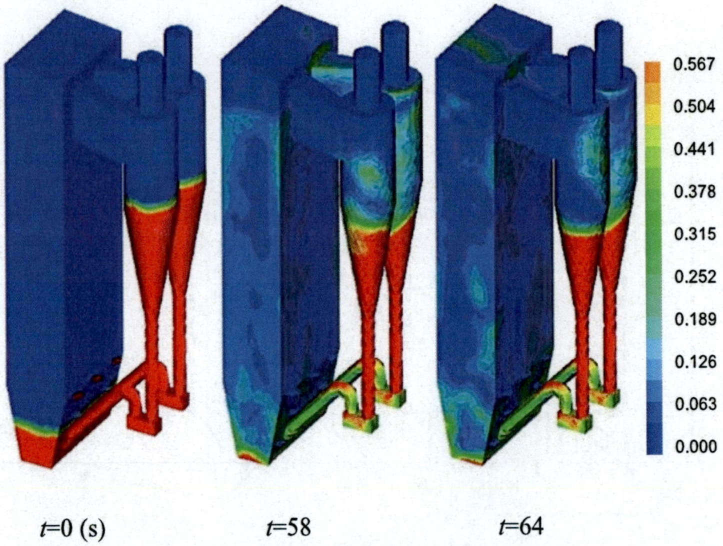


**Fig. 6.2** Instantaneous concentrations of particles in a CFB combustor

equal. It can be seen that there are much more particles near the wall than in the centre, especially near the wall close to the cyclone side.

Figure 6.3 shows the distribution of averaged velocities of gas phase and particles at three different heights. The velocities of gas phase and particles are higher in the center regime than those near the walls. The velocity of particles is negative at the east and west walls and the front wall, which means that particles flow down close to the walls. However, it is positive at the rear wall, which indicates that particles flow up. The particles in the separator and dip leg return to the furnace from a loop seal by secondary air. Thus, particles near the west wall flow up with secondary air. Therefore, the secondary air influences the motion of particles in the furnace.

The predicted pressure distribution along height in the CFB combustor is shown in Fig. 6.4. The high pressure drops appears near the bottom. The pressure drop gradient is much lower in the upper part than that in the lower part of the furnace.

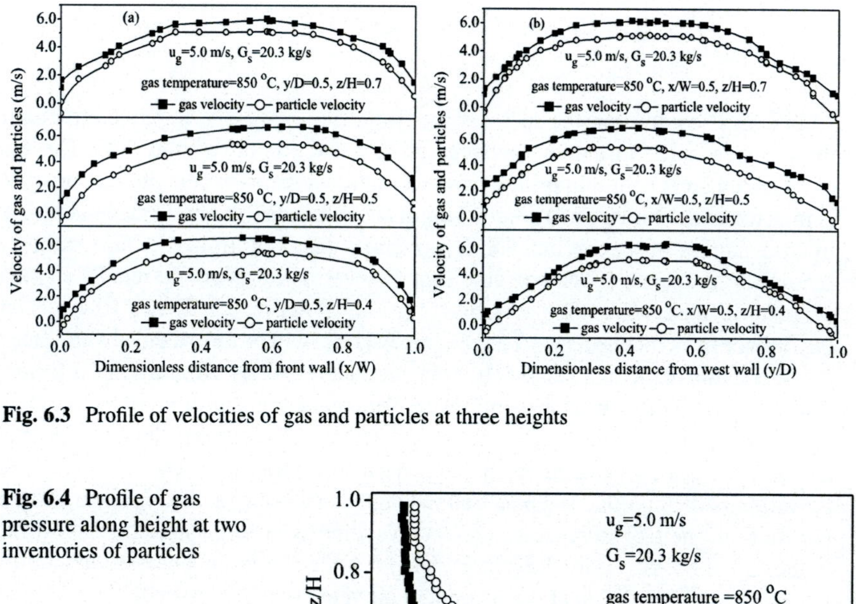


**Fig. 6.3** Profile of velocities of gas and particles at three heights

**Fig. 6.4** Profile of gas pressure along height at two inventories of particles

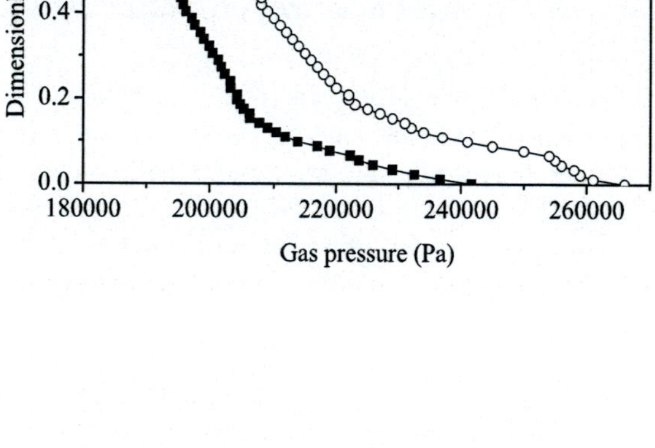

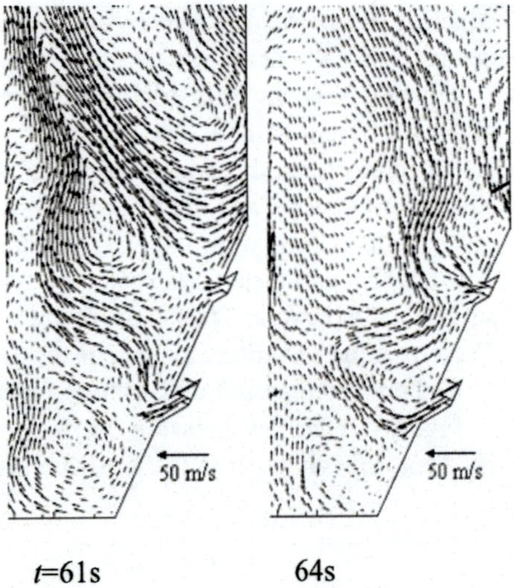


**Fig. 6.5** Instantaneous gas velocities at the local secondary air ports

In a CFB combustor, the secondary air injection is mainly employed to control the nitrogen oxide formation, enhance mixing and stimulate solid flow. The total combustion air is split into primary and secondary streams, where the secondary air is injected to the furnace at a certain height above the distributor. Understanding the hydrodynamics of the gas and solids movement resulting from the introduction of secondary air can have considerable significance in improving design. The typical local gas jet observed in the simulation is presented in Fig. 6.5. The air jet ejects into the furnace in an incline angle. The secondary air jet will be turned on upwards due to the interaction of gas and solids. The local circulation of gas with particles is formed. Thus, the mixing of particles is improved. Due to the transfer of momentum, the jet cross section increases with the decrease of the gas velocity along the jet to maintain the constant gas flow rate inside the jet, and the penetration length is varied. There are different definitions of jet penetration length used in the literature. The maximum penetration length of secondary air jet is considered to be the distance from the orifice to the end of the curvature. Results indicate that the maximum penetration length is in the range of 0.5–1.3 m at the secondary air velocity of 50–90 m/s.

### 6.1.2 *Hybrid Pulverized-Fluidized Bed Coal Combustors*

In coal power plants, the coal is stored at silos, and it is pulverized by mills to improve its thermal use before entering in the combustion chamber. The pulverized coal (PC) is then dragged by primary air and fed into the burners. Secondary air is injected in the burners to promote the dispersion of the coal inside of the reaction chamber, seeing Fig. 6.6. The burners are positioned at the corners of the combustion chamber. The high velocity jets of pulverized coal and primary air are injected at the center of

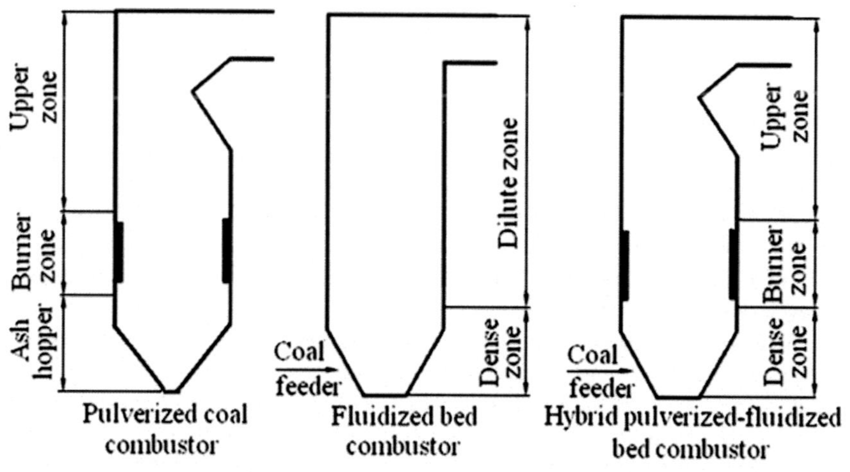


**Fig. 6.6** Schematic representations of pulverized coal, fluidized bed and hybrid pulverized-fluidized bed combustors

the furnace to induce strong circulation of gas and particles. The secondary air jets are added to promote the intensity of tangential circle formed in the center of the furnace. Water is heated to produce superheated, high pressure steam through tubes lined in the furnace and superheaters. The high pressure steam is passed into a steam turbine to generate electricity. The flue gases pass through scrubbers before they exit a plant's smokestack. The main advantages of pulverized coal combustors are high combustion efficiencies of coal particles. On the other hand, the main disadvantages of pulverized coal combustors are the plant still requires the high efficiency $NO_x$ removal devices to control $NO_x$ emissions using selective catalytic reduction (SCR) and selective noncatalytic reduction (SNCR) because PC coal combustion leads to emission of $SO_2$ and $NO_x$ pollutants. The flue gas desulfurization (FGD) is viewed as a major and effective measure to control $SO_2$ emission. PC still limits to coal with relatively low ash content of power plants.

Fluidized bed coal combustion system is a bed or layer of small solid particles, normally with diameters of less than 6.0 mm at the bottom of a combustion chamber. Fluidizing air is blown into the chamber through air distributor. This air has sufficient velocity that as it passes up through particles it entrains them so that they become suspended in the dense zone near the bottom of the combustion chamber. This suspension of solid particles displays many of the properties of a fluid, so that the particles move and jostle one another. The small coal particles also are suspended in the dilute zone of the upper part of the furnace. Combustion takes place at low temperatures from 800 to 900 °C in the dense and dilute zones along the furnace height, resulting in reduced $NO_x$ formation compared with pulverized coal combustion process. Sulfur dioxide emissions can be reduced by the injection of limestone into the bed and subsequent removal of ash together with reacted sorbent without the need for flue gas desulphurization. Fluidized bed systems can reduce sulfur dioxide by 90–95% and nitrogen oxides by 90% or more. Further, fluidized bed combustion provides a more uniform temperature in the combustion zone, making heat capture and transfer possible with smaller capture surfaces and this can, in principle, reduce the overall size and cost for a given heat output. The fluidized bed combustion is principally of value for low grade, high ash coals which are difficult to pulverized coal combustors. It is also suitable for co-firing coal with low grade fuels, including waste materials. On the other hand, the disadvantages of fluidized bed combustion include low combustion efficiency because the coal particle size is much larger than that in a pulverized coal plant and it operates at a low temperature in comparison to pulverized coal combustion chamber.

A hybrid pulverized-fluidized bed coal combustor is proposed with a fluidized dense zone near the bottom and pulverized coal combustion at the burner zone and suspension burning zone at the upper part of the furnace. The fluidized bed coal particles enter into the dense zone, and they are fluidized by fluidizing air from the distributor at the bottom. The pulverized coal particles enter through the burners by means of primary air and secondary air. The finer particles from fluidized coal particles and pulverized coal particles are further burned in the upper part of the furnace. The advantages of the hybrid pulverized-fluidized bed coal combustor are as follows: (1) high combustion efficiency because the two traverse flows from the high velocity jets of pulverized coal and the fluidized coal particles collides to promote turbulence and retention of the coal particles by an enough time to accomplish the combustion in suspension. (2) It is also useful for coal-biomass combustion or gasification. The fluidized biomass is burned at the bottom, and the suspended pulverized coal combusts at the burner zone. Another important behavior is that the boiler responsiveness is improved to load demand changes, which is very important for compensating the unpredictable load balance changes of the electricity system.

The combustion of coal particles undergoes a series of processes in hybrid pulverized- fluidized bed coal combustor: initially drying and devolatilization, subsequently oxidation of volatiles, and finally char combustion reactions (Shuai et al. 2014a). For simplicity, the moisture of the fuel is added to the volatiles and the drying and devolatilization parallel take place. The volatile consists of $CH_4$, CO, $H_2O$, $CO_2$, $H_2$ and $C_2H_6$. In the coal devolatilization process, we consider that the raw coal decomposed into char in the solids phase and volatile matter in the gas phase. This process is modeled by the following reaction:

$$1\ \text{kg dry coal} \rightarrow \text{Y}_\text{v}\ \text{kg volatiles} + \text{Y}_\text{c}\ \text{kg char} + (1 - \text{Y}_\text{v} - \text{Y}_\text{c})\ \text{kg ash} \quad \text{(R1)}$$

The corresponding reaction rate is given in Table 6.2, where $A_v$ is a pre-exponential factor, $E_v$ is the activation energy and $R$ is the universal gas constant. The constants $A_v$ and $E_v$ depend on coal type, and they are $2.02 \times 10^3$ and $3.11 \times 10^7$, respectively.

The volatile combustion takes place according to the homogeneous reactions, including methane oxidation (R2), ethane reaction with oxygen (R3), complete combustion (R4) and hydrogen oxidation (R5).

**Table 6.2** Homogeneous reaction rates of coal particles

| | Reaction rate (mol m$^{-3}$ s$^{-1}$) | Reaction rate constant |
|---|---|---|
| (R1) | $R_1 = k_1 Y_c$ | $k_1 = 2.02 \times 10^3 \exp\left(\frac{-3.11\times 10^7}{RT_g}\right)$ |
| (R2) | $R_2 = k_2\left(\frac{Y_{O_2}\rho_g}{M_{O_2}}\right)^{0.8}\left(\frac{Y_{CH_4}\rho_g}{M_{CH_4}}\right)^{0.7}$ | $k_2 = 1.58 \times 10^{10} \exp\left(\frac{-24343}{T_g}\right)$ |
| (R3) | $R_3 = k_3\left(\frac{Y_{C_2H_6}\rho_g}{M_{C_2H_6}}\right)\left(\frac{Y_{O_2}\rho_g}{M_{O_2}}\right)$ | $k_3 = 1.585 \times 10^{10} \exp\left(\frac{-24157}{T_g}\right)$ |
| (R4) | $R_4 = k_4 Y_{CO} Y_{H_2O}^{0.5} \frac{17.5 Y_{O_2}}{1+24Y_{O_2}}\left(\frac{P}{RT_g}\right)^{1.8}$ | $k_4 = 3 \times 10^{10} \exp\left(\frac{-6.699\times 10^7}{RT_g}\right)$ |
| (R5) | $R_5 = k_5\left(\frac{Y_{O_2}\rho_g}{M_{O_2}}\right)\left(\frac{Y_{H_2}\rho_g}{M_{H_2}}\right)^{1.5}$ | $k_5 = 1.63 \times 10^9 T_g^{1.5} \exp\left(\frac{-3420}{T_g}\right)$ |

$$CH_4 + \frac{3}{2}O_2 \rightarrow CO + 2H_2O \tag{R2}$$

$$C_2H_6 + \frac{5}{2}O_2 \rightarrow 2CO + 3H_2O \tag{R3}$$

$$CO + \frac{1}{2}O_2 \rightarrow CO_2 \tag{R4}$$

$$H_2 + \frac{1}{2}O_2 \rightarrow H_2O \tag{R5}$$

The solids phase is a mixture consisting of coal, char and ash particles with limestone particles. The heterogeneous combustion of char is modeled with consideration of a mechanism factor which describes the split of the combustion products carbon monoxide CO and carbon dioxide $CO_2$ (Yunhau et al. 2006). The char is reacted according to heterogeneous combustion (R6) and Boudouard reaction (R7).

$$C(s) + \Phi O_2(g) \rightarrow 2(1-\Phi)CO(g) + (2\Phi - 1)CO_2(g) \tag{R6}$$

$$C(s) + CO_2(g) \rightarrow 2CO(g) \tag{R7}$$

The mechanism factor $\Phi$ is determined by particles diameter and combustion temperature as well the ratio of produced carbon monoxide to carbon dioxide.

$$\Phi = \begin{cases} (2+\kappa)/(2\kappa+2) & d_p < 0.05 \text{ mm} \\ \frac{(\kappa+2)}{(2\kappa+2)-(p/0.095)(100d_p-0.005)} & 0.005 \le d_p \le 1.0 \text{ mm} \\ 1.0 & d_p > 1.0 \text{ mm} \end{cases} \tag{6.1}$$

$$\kappa = \frac{Y_{CO}}{Y_{CO_2}} = 2500\exp(-\frac{6240}{T_s}) \tag{6.2}$$

**Table 6.3** Heterogeneous reaction rates of coal and limestone particles

| | Reaction rate (mol $m^{-3}$ $s^{-1}$) | Reaction rate constant |
|---|---|---|
| (R6) | $R_6 = \frac{6\alpha_s\rho_s Y_c}{d_p\rho_c}k_6 Y_{O_2}$ | $k_6 = \frac{RT_s/M_c}{(1/k_{cr})+(1/k_{cd})}$, $k_{cr} = 8910\exp\left(\frac{-1.4947\times10^8}{RT_s}\right)$, $k_{cd} = \frac{Sh\varphi\left(D_g+\frac{\mu_g}{\rho_g Sc_t}\right)M_c}{d_{s1}R_g T_g}$ |
| (R7) | $R_7 = \frac{k_7 Y_{CO_2} Y_c}{1+k_{7,co_2}Y_{CO_2}+k_{7,co}Y_{CO}}$ | $k_7 = 3.1785\times10^{10}\exp\left(\frac{-2.68\times10^8}{RT_s}\right)$ $k_{7,co_2} = 66\exp\left(\frac{-2.55\times10^7}{RT_s}\right)$, $k_{7,co} = 120\exp\left(\frac{-2.55\times10^7}{RT_s}\right)$ |
| (R8) | $R_8 = k_8\alpha_s\rho_s Y_{CaCO_3}S_{CaCO_3}\frac{p_e-p_{CO_2}}{p_e}$ | $k_8 = 6.078\times10^4\exp\left(\frac{-2.05\times10^8}{RT_s}\right)$ $p_e = 4.192\times10^{12}\exp\left(\frac{-1.702\times10^8}{RT_g}\right)$, $S_{CaCO_3} = 1.26m^2g^{-1}$ |
| (R9) | $R_9 = k_9\alpha_s\rho_s Y_{CaO}s_g Y_{SO_2}a$ | $k_9 = 490\exp\left(\frac{-1.75\times10^7}{RT_s}\right)$ $a = \exp\left[\frac{-571Y_{CaSO_4}}{(Y_{CaCO_3}+Y_{CaO}+Y_{CaSO_4})M_{CaCO_3}}\right]$ $s_g = \begin{cases}-38.4T_s+5.6\times10^4 & T_s\geq1253K\\ 35.9T_s-3.67\times10^4 & T_s<1253K\end{cases}$ |
| (R10) | $R_{10} = k_{10}\alpha_s\rho_s Y_{CaCO_3}S_{CaCO_3}Y_{SO_2}$ | $k_{10} = 0.72\exp\left(\frac{-6.4\times10^7}{RT_s}\right)$ |

The chemical reactions are listed in Table 6.3, were $T$ is temperature. $Y_c$, $\rho_c$ and $M_c$ are the mass fraction, density and molar weight of char particles, and $Y_{O2}$, $Y_{CO2}$ and $Y_{CO}$ are the molar fractions of oxygen, carbon dioxide and carbon monoxide.

In the desulphurization process, the particles of CaO are produced through calcination from the limestone and used for capturing $SO_2$. The main reactions are considered as follows:

$$CaCO_3 \rightarrow CaO + CO_2 \tag{R8}$$

$$CaO + SO_2 + \frac{1}{2}O_2 \rightarrow CaSO_4 \tag{R9}$$

$$CaCO_3 + SO_2 + \frac{1}{2}O_2 \rightarrow CaSO_4 + CO_2 \tag{R10}$$

The Limestone calcination (R8) generally takes place at temperatures of 800–900 °C, and the reaction between CaO and $SO_2$ (R9) occurs at more significant speed between 600 and 1000 °C. The injection of limestone up to temperatures higher than 1050 °C is not appropriate with respect to the deactivation of the calcine created (so-called overfiring) as a consequence of a change of the porous structure. In simulations of sulfur capture, the sulfur in the coal is assumed to be released during devolatilization and totally oxidized to sulphur dioxide $SO_2$. A fraction of sulphur dioxide is discharged with gas phase and the rest is captured due to the addition of limestone particles. In numerical simulations, the limestone particles consist of three pseudo-species ($CaCO_3$, CaO and $CaSO_4$). The weights fraction of $CaCO_3$ and CaO are 0.9 and 0.1, respectively.

The production of $NO_x$ during coal combustion is released as volatile and partially retained in the char combustion in part dependent on the chemical characteristics of the coal, but particularly, on the firing conditions within the combustion chamber. The most important factors in coal firing that influence $NO_x$ formation are the flue gases associated with the oxygen concentration and temperature (Shuyan et al. 2008a). The reaction rates are given in Table 6.4 as a function of gas temperature and species concentration.

$$NO + Char(s) \rightarrow 0.5N_2 + CO \tag{R11}$$

$$NH_3 + 1.25O_2 \rightarrow NO + 1.5H_2O \tag{R12}$$

$$NH_3 + 0.75O_2 \rightarrow 0.5N_2 + 1.5H_2O \tag{R13}$$

$$NO + CO \rightarrow 0.5N_2 + CO_2 \tag{R14}$$

$$HCN + 0.5O_2 \rightarrow CNO \tag{R15}$$

$$CNO + 0.5O_2 \rightarrow NO + CO \tag{R16}$$

$$CNO + NO \rightarrow N_2O + CO \tag{R17}$$

$$N_2O + CO \rightarrow N_2 + CO_2 \tag{R18}$$

$$N_2O + Char(s) \rightarrow N_2 + CO \tag{R19}$$

$$NH_3 + 0.5O_2 + NO \rightarrow N_2 + 1.5H_2O \tag{R20}$$

$$2N_2O \rightarrow 2NO + N_2 \tag{R21}$$

**Table 6.4** $NO_x$ chemical reaction rates

| Chemical reactions | Reaction rate R ($kmol/(m^3 \cdot s)$) | Reaction rate constant |
|---|---|---|
| (R11) | $R_{11} = \alpha_g k_{11} N \pi d_p^2 Y_{NO}$ | $k_{11} = 5.85 \times 10^4 \exp(-99768/RT)$ |
| (R12) | $R_{12} = \alpha_g k_{12} Y_{NH_3} Y_{O_2}$ | $k_{12} = 2.73 \times 10^{11} \exp(-3.17 \times 10^5/RT)$ |
| (R13) | $R_{13} = \frac{k_{13} Y_{NH_3} Y_{O_2}}{Y_{O_2} + 0.054}$ | $k_{13} = 3.38 \times 10^4 \exp(-83140/RT)$ |
| (R14) | $R_{14} = \alpha_g k_{14} \frac{(k_{141} Y_{NO}(k_{142} Y_{CO} + k_{143}))}{(k_{141} Y_{NO} + k_{142} Y_{CO} + k_{143})}$ | $k_{14} = 1.952 \times 10^7 \exp(-1.58 \times 10^5/RT) k_{141} = 0.1826; k_{142} = 0.00786; k_{143} = 0.002531$ |
| (R15) | $R_{15} = \alpha_g k_{15} Y_{HCN} Y_{O_2}$ | $k_{15} = 214 \times \exp(-83140/RT)$ |
| (R16) | $R_{16} = \alpha_g k_{16} Y_{HCN} Y_{O_2} \left( \frac{k_{161}}{k_{161} + k_{162} Y_{NO}} \right)$ | $k_{16} = 214 \exp(-83140/RT) k_{161}/k_{162} = 1.02 \times 10^6 \exp(-2.12 \times 10^5/RT)$ |
| (R17) | $R_{17} = \alpha_g k_{17} Y_{HCN} Y_{O_2} \left( \frac{k_{172} Y_{NO}}{k_{171} + k_{172} Y_{NO}} \right)$ | $k_{17} = 214 \exp(-83140/RT) k_{171}/k_{172} = 1.02 \times 10^6 \exp(-2.12 \times 10^5/RT)$ |
| (R18) | $R_{18} = \alpha_g k_{18} Y_{N_2O} Y_{CO}$ | $k_{18} = 5.01 \times 10^{10} \exp(-4.40 \times 10^4/RT)$ |
| (R19) | $R_{19} = \alpha_g k_{19} N \pi d_p^2 Y_{N_2O}$ | $k_{19} = 2.9 \times 10^6 \exp(-1.41 \times 10^5/RT)$ |
| (R20) | $R_{20} = \alpha_g k_{20} Y_{NH_3}^{0.5} Y_{NO}^{0.5} Y_{O_2}$ | $k_{20} = 1.1 \times 10^{15} \exp(-2.3 \times 10^5/RT)$ |
| (R21) | $R_{21} = \alpha_g k_{21} Y_{N_2O}$ | $k_{21} = 1.75 \times 10^8 \exp(-1.98 \times 10^5/RT)$ |

The combustion chamber is comprised of a continuous upflowing gas phase with dispersed solids along the heat transfer surfaces. The total heat transfer coefficient $h_t$ may be written as the sum of the convective and radiative heat transfer coefficients. The Nusselt number for the individual particles is determined for a porosity $\alpha_f$ in the range of 0.35–1.0

$$Nu = (7 - 10\alpha_f + 5\alpha_f^2)(1 + 0.7Re^{0.2}Pr^{1/3}) + (1.33 - 2.4\alpha_f + 1.2\alpha_f^2)Re^{0.7}Pr^{1/3} \tag{6.3}$$

where $Re$ and $Pr$ are the Reynolds and Prandtl numbers for the gas phase, respectively.

The radiation is the dominant energy transport mechanism to surrounding heating surfaces in the furnace. It is also one of the most complex problems since it involves solving an integral– differential equation for radiation heat transfer. The two most commonly used radiation models are P1 model and Discrete Ordinates (DO) model.

The advantage of the P1 model over the DO model is its small numerical cost. Moreover, the P1 model showed low accuracy when complex geometries were involved. Radiation is modeled using a P1 method for a gray, absorbing, emitting and scattering gas medium containing absorbing, emitting and isotropically scattering solid particles. The transport equation for the incident radiation $G$ is

$$\nabla \cdot (\Gamma \nabla G) + 4\pi \left(a\frac{\sigma T^4}{\pi} + E_p\right) - (a + a_p)G = 0 \tag{6.4}$$

$$\Gamma = \frac{1}{3(a + a_p + \sigma_p)} \tag{6.5}$$

where $a$ is the absorption coefficient for the gray gas and σ is the Stefan-Boltzmann constant. $E_p$ and $a_p$ are the equivalent emission of particles and the equivalent absorption coefficient of particles, respectively. Here, the equivalent radiative properties refer to the cell-mean radiative properties. For a computational cell containing $N$ particles component with $N_n$ particles numbers.

$$E_p = \lim_{V \to 0} \sum_{n=1}^{N} \varepsilon_{pn} \frac{d_{pn}^2 N_n \sigma T_{pn}^4}{4V} \tag{6.6}$$

$$a_p = \lim_{V \to 0} \sum_{n=1}^{N} \varepsilon_{pn} \frac{\pi N_n d_{pn}^2}{4V} \tag{6.7}$$

$$\sigma_p = \lim_{V \to 0} \sum_{n=1}^{N} (1 - f_{pn})(1 - \varepsilon_{pn}) \frac{\pi N_n d_{pn}^2}{4V} \tag{6.8}$$

where $\varepsilon_{pn}$ is the equivalent emittance for particles component $n$. $f_{\mathrm{pn}}$ is the scattering factor associated with the nth particles component. These properties are used to solve for the incident radiation $G$. Once the field of incident radiation $G$ is obtained, the radiation source term in the energy equation is obtained as

$$-Q_{red} = aG - 4a\sigma T^4 \tag{6.9}$$

The numerical simulations are performed for a tangential corner pulverized coal-fired combustor, fluidized bed and hybrid pulverized-fluidized bed combustors. The pulverized coal combustor has a maximum continuous rating (MCR) of 127.8 kg/s. The finer pulverized coal particles and combustion air is fed via burner system, showing in Fig. 6.7. The total air split into primary air and secondary air. The burner zone is made of three unit burners. The over fire air (OFA) is arranged at the top of burner system. Partial combustion air is injected through OFA nozzles. The fluidized bed combustor has a MCR of 125.0 kg/s. The mid and coarse coal particles are fed by means of screw feeder at the front wall. The total air split into fluidizing air from

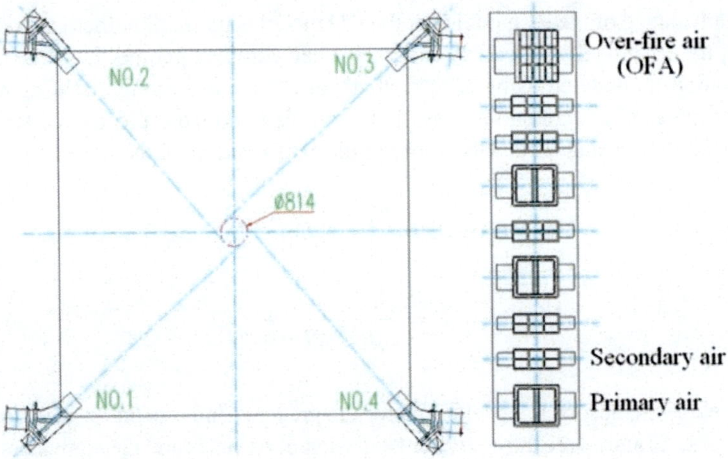


**Fig. 6.7** Arrangement of combustion burners

the distributor and secondary air at the dense zone. The hybrid pulverized-fluidized bed coal boiler has a MCR of 127.8 kg/s. The total air split into fluidizing air from the distributor, primary air and secondary air from the burner system. Fuel also split into the fluidized bed coal with the mid and coarse particles through the screw feeder and finer pulverized coal particles from the burner system. The main parameters are listed in Table 6.5. Proximate and ultimate analysis of the coal is presented in Table 6.6.

Figure 6.8 shows gas temperature in the hybrid pulverized-fluidized bed coal furnace, and pulverized coal and fluidized bed furnaces. In the hybrid pulverized-fluidized bed coal furnace, the uniform gas temperature distribution exists in the dense zone at the bottom of the furnace. The gas temperature rises at the burner zone and the gradually decreases along height. On the other hand, the gas temperature is very low in the ash hopper of the bottom of the furnace, and reach maximum at the burner zone. It decreases at the upper zone of the furnace. In the fluidized bed furnace, the uniform gas temperature appears, and rises at the inlet of the dilute zone. After that the gas temperature slightly decreases along the height of the dilute zone at the upper part of the furnace.

Figure 6.9 shows the gas velocities of the first primary air and second air burners layer in the hybrid pulverized-fluidized bed furnace at the air fluidizing velocity of 1.3 m/s. The primary air jet with pulverized coal particles ejecting into the center of the furnace suffer from the interaction of vertical fluidizing air from the dense zone. The primary air has two principle objectives, namely pulverized coal combustion and formation of a circle at the furnace center. Simulated results show that both circles of primary air and secondary air burners are generated, indicating the vertical fluidizing gas from the dense zone does not effect on hydrodynamics of burner system. The formed tangential circles enhance burning rate of pulverized coal particles.

Figure 6.10 shows the distributions of NO gas species concentration in the hybrid

**Table 6.5** Parameters used in simulations of three different furnaces

| Parameter | unit | Hybrid pulverized-fluidized furnace | Pulverized coal furnace | Fluidized bed furnace |
|---|---|---|---|---|
| Furnace height | m | 28.67 | 28.67 | 36.85 |
| Furnace width × depth | m | 10.38 × 10.38 | 10.38 × 10.38 | 15.32 × 7.22 |
| Distributor width × depth | | 3.6 × 10.38 | / | 3.6 × 10.38 |
| Diameter of pulverized coal particles | μm | 50 | 50 | / |
| Diameter of fluidized mid coal particles | mm | 0.1 | / | 0.1 |
| Diameter of fluidized coarse coal particles | mm | 1.5 | / | 1.5 |
| Density of coal particles | $kg/m^3$ | 2000 | 2000 | 2000 |
| Primary air velocity | m/s | 25.0 | 25.0 | / |
| Secondary air velocity | m/s | 45.0 | 45.0 | 55.0 |
| Fluidizing gas velocity | m/s | 1.3 | / | 1.3 |
| Fluidized bed coal mass flux | kg/s | 25.4 | / | 91.6 |
| Pulverized coal mass flux | kg/s | 22.9 | 91.6 | / |
| Static bed height of dense zone | m | 1.2 | / | 1.2 |
| Equivalence ratio (SR) | / | 1.1 | 1.1 | 1.1 |
| Ca/S mole ratio | / | 1.5 | / | 1.5 |

**Table 6.6** Ultimate and proximate analyses of coal

| Ultimate analysis | | Proximate analysis | |
|---|---|---|---|
| $C_{ad}$ | 40.53 wt % | Moisture, $M_{ad}$ | 10.6 wt % |
| $H_{ad}$ | 2.94 wt % | Ash, $A_{ad}$ | 35.00 wt % |
| $N_{ad}$ | 0.55 wt % | Fixed carbon, FC | 25.34 wt % |
| $O_{ad}$ | 9.80 wt % | Volatile, $V_{ad}$ | 43.96 wt % |
| $S_{ad}$ | 0.58 wt % | $Q_{net,ar}$ | 15,900 kJ/kg |

**Table 6.7** Mathematical model of chemical reactions in CLC system

| Reactions | Reaction rates |
|---|---|
| (FR-R1) | $R_1 = 0.2\varepsilon_s Y_{CH_4}^{0.2} \exp[-\frac{5000}{RT_g}][3(1-X)^{2/3}]$ |
| (FR-R2) | $R_2 = 0.15\varepsilon_s Y_{H_2}^{0.4} \exp[-\frac{5000}{RT_g}][3(1-X)^{2/3}]$ |
| (FR-R3) | $R_3 = 0.059\varepsilon_s Y_{CO}^{0.6} \exp[-\frac{5000}{RT_g}][3(1-X)^{2/3}]$ |
| (FR-R4) | $R_4 = 0.84\varepsilon_s Y_{O_2}^{0.7} \exp[-\frac{22000}{RT_g}][3(1-X)^{2/3}]$ |
| (AR-R1) | $R_5 = 0.8677 Y_{O_2}^{0.958} \exp(-\frac{43390}{RT_g})[3(1-X)^{2/3}]$ |

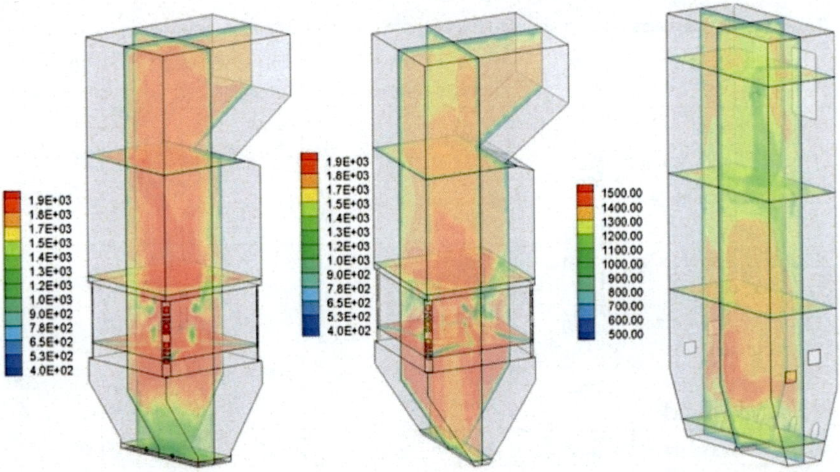


**Fig. 6.8** Gas temperatures of hybrid pulverized-fluidized bed (left), pulverized coal (middle) and fluidized bed (right) furnaces

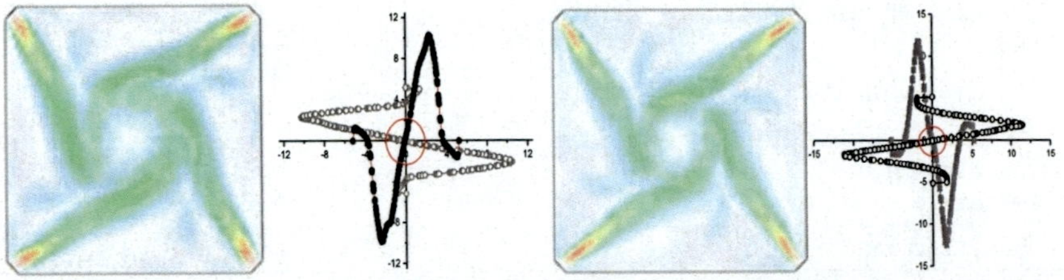

**Fig. 6.9** Gas velocities of primary air (left) and secondary air (right) in the first burner layer of hybrid pulverized-fluidized bed furnace

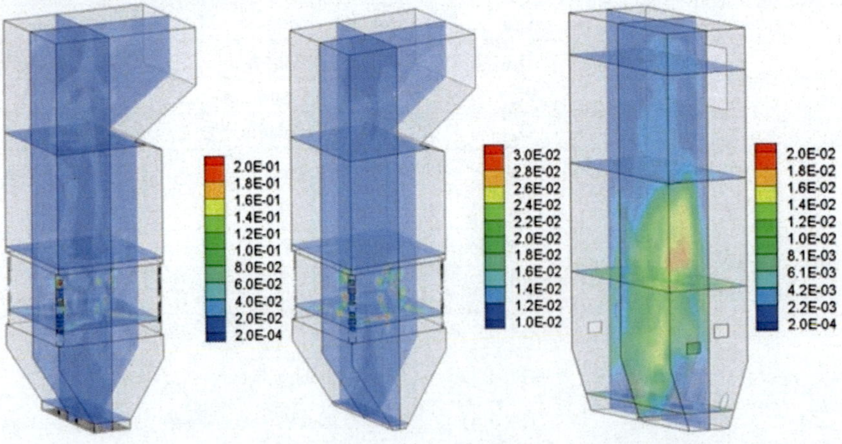


**Fig. 6.10** Distributions of NO concentration of hybrid pulverized-fluidized bed (left), pulverized coal (middle) and fluidized bed (right) furnaces

pulverized-fluidized bed, pulverized coal and fluidized bed furnaces. In the hybrid pulverized-fluidized bed and fluidized bed furnaces, the fluidizing air has two objectives of combustion and fluidization. In case of combustion, increasing fluidizing air increases bed temperature when the fluidized bed coal fuel feed rate is proportionally increased. Increased fluidizing air decreases the bed temperature at a constant fuel feed rate as the heat is carried by the flue gas away from the dense phase. Both bed temperatures are in the range of 860–980 °C. The low gas temperature leads to low NO gas species concentration. The high NO gas species concentrations are found at the burner zone in the hybrid pulverized-fluidized bed and pulverized coal furnaces because of high gas temperature. The thermal $NO_x$ is formed at temperatures exceeding 1540 °C. Since gas temperatures in the hybrid pulverized-fluidized bed furnace are well below this temperature, thermal $NO_x$ is not a major component of its nitrogen oxide emissions. $NO_x$ emissions from hybrid pulverized-fluidized bed and fluidized bed furnaces are lower than those from pulverized coal furnace.

The distributions of $SO_2$ gas species concentration are shown in Fig. 6.11 in the hybrid pulverized-fluidized bed, pulverized coal and fluidized bed furnaces. The simulated $SO_2$ gas species concentrations are low at the dense zone of hybrid pulverized-fluidized bed and fluidized bed furnaces. It is high at the burner zone of the hybrid pulverized-fluidized bed and pulverized coal furnaces. The conversion rate to $CaSO_4$ decreases with increase in temperature as the calcium oxide pores are plugged at a faster rate by higher molar volume $CaSO_4$. The reactivity of sorbent increases with a rise in combustor temperature and is optimum in the temperature range of 800–850 °C, improving efficiency of sulfur capture in the dense zone of hybrid pulverized-fluidized bed and fluidized bed furnaces.

The heat is mainly extracted via the furnace walls through convection and radiation in the furnace. The distributions of heat fluxes are shown in Fig. 6.12 in the hybrid pulverized-fluidized bed, pulverized coal and fluidized bed furnaces. In the pulverize

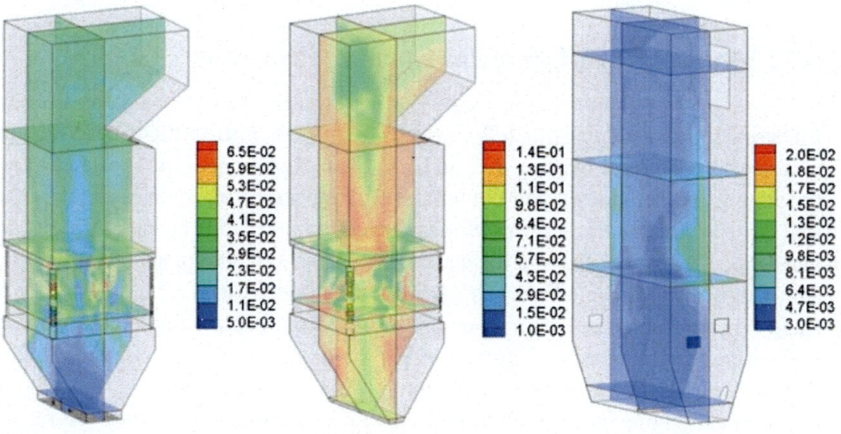


**Fig. 6.11** Distributions of $SO_2$ concentration of hybrid pulverized-fluidized bed (left), pulverized coal (middle) and fluidized bed (right) furnaces

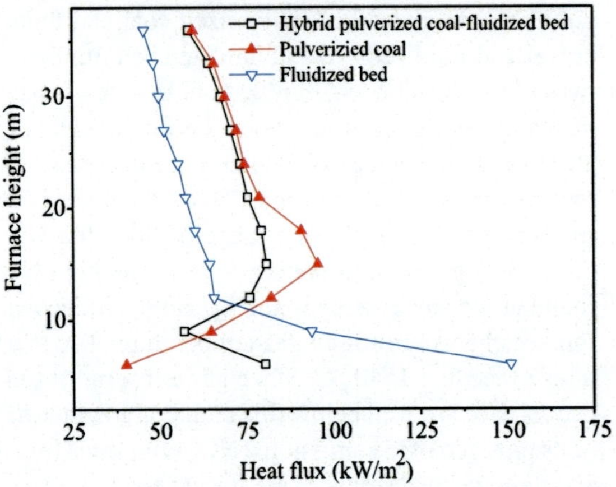


**Fig. 6.12** Distributions of heat fluxes are shown in the hybrid pulverized-fluidized bed, pulverized coal and fluidized bed furnaces

coal furnace, the heat flux increases from the ash hopper, reach maximum at the burner zone, and then decrease in the upper part along the furnace height. The heat flux has the largest value in the burner zone, and the lowest one at the ash hopper. In the fluidized bed furnace, the heat flux reduces along furnace height. The lower heat fluxes are found in the upper dilute region in the fluidized bed furnace. On the other hand, the heat flux decreases in the dense zone, then increases in the burner zone in the hybrid pulverized- fluidized bed furnace, improved the ignition stability of pulverized coal at the burner zone. The heat flux decreases in the upper region of the furnace. The ratio of the maximum heat fluxes between the pulverized coal furnace and the hybrid pulverized-fluidized bed furnace is 1.2, indicating that the heat flux is reduced at the burner zone in the hybrid pulverized-fluidized bed furnace, reducing the risk of slagging in the area close to the furnace wall.

## 6.2 Fluidized Bed Chemical-Looping Processes

Chemical looping combustion (CLC) is a technique that utilizes metal oxides instead of air in the process of burning fuel. In CLC, the carbon-containing fuel such as coal or syngas is burned not with air but with a solid oxygen carrier such as a metal oxide, $M_xO_y$, to produce a nitrogen-free effluent gas containing a high concentration of $CO_2$ in a form, after condensing out the water component, suitable for subsequent capture and storage. The depleted oxygen carrier is then regenerated with air in a second stage. The two processes are conveniently carried out in two coupled fluidized-bed fuel reactor and air reactor between which the oxygen carrier is circulated, where they provide the oxygen necessary for combustion. Since only pure oxygen is used in the combustion process, and no nitrogen oxide is produced. CLC involved reaction of a gaseous hydrocarbon with a metal oxide that is taken from taken from the highest state to the lowest one of oxidation, following the global reaction scheme:

$$(2n + m)M_yO_x + C_nH_{2m} = (2n + m)M_yO_{x-1} + mH_2O + nCO_2 \quad (6.10)$$

Alternatively, if a reforming step is being considered, the amount of oxidant is reduced to allow for the production of CO and $H_2$:

$$nM_yO_x + C_nH_{2m} = nM_yO_{x-1} + mH_2 + nCO \quad (6.11)$$

In the case of combustion, after the water produced from hydrocarbon oxidation is removed, the flue gas stream contains an effectively pure stream of $CO_2$ suitable for sequestration. The metal oxide is regenerated in a separate reactor by reaction with air, and the regenerated solid can be transferred back for further reaction with the fuel gas:

$$M_yO_{x-1} + \text{Air} = M_yO_x \quad (6.12)$$

Typically, the oxidation reaction is strongly exothermic, so that overall, the system yields the heating value of the fuel. Such a cycle represents an elegant way of oxidizing a fuel gas by effectively achieving air separation, without using cryogenic or membrane technology, and, avoiding or minimizing the formation of fuel-$NO_x$. In practice, these cycles typically operate at temperatures in the range of 800–1200 °C, which normally ensures that the reactions occur at a sufficient rate to be compatible with fluidized bed systems.

### *6.2.1 Diameter Transformed Fluidized Bed Chemical-Looping Combustion (CLC)*

The use of natural gas and/or syngas has been demonstrated in fluidized bed CLC applications. The CLC system consists of an air reactor (AR) and a fuel reactor (FR), as shown in Fig. 6.13a. The metal oxide particles are used as oxygen carrier that circulates between the two reactors (Bougamra and Huilin 2014; Shuai et al. 2014b). The AR is a diameter transformed fluidized bed which several reaction processes have been developed. The oxygen carrier particles are transported together with the air stream to the top of the AR, where they are then transferred to the fuel reactor (FR) using a cyclone (CY). The FR is a bubbling fluidized bed reactor where oxygen carrier particles react with hydrocarbon fuel and get reduction. The reduced oxygen carrier particles are transported back to the AR through the loop seal (LS) where they react with oxygen in the air and are oxidized back to metal oxide. The exhaust from the FR mainly consists of $CO_2$ and water vapor. After condensation of the water in the exit gas from the FR, the remaining $CO_2$ gas is compressed and cooled to yield liquid $CO_2$, which can be disposed in various ways.

The oxygen carrier used in the investigation here is NiO particles. The reduction in the FR and oxidation in the AR of the oxygen carrier are described by the

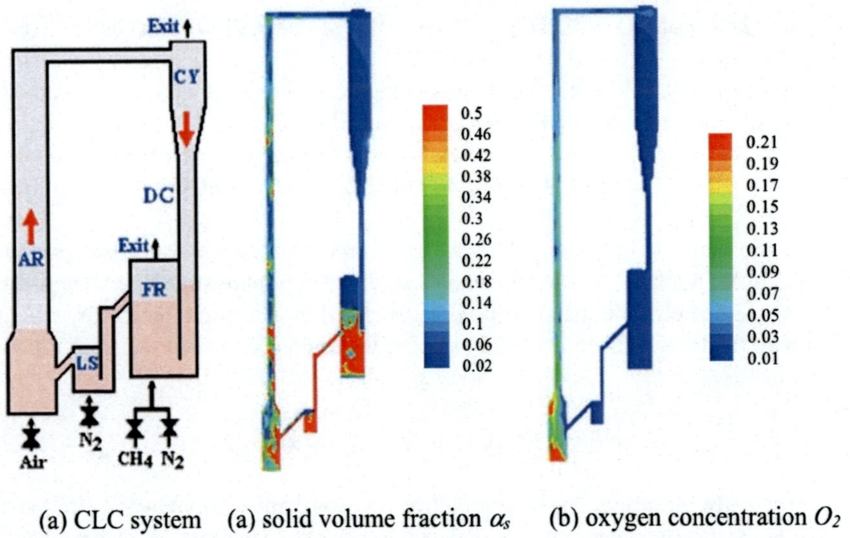


**Fig. 6.13** Instantaneous solids volume fraction and volume fraction of oxygen gas species

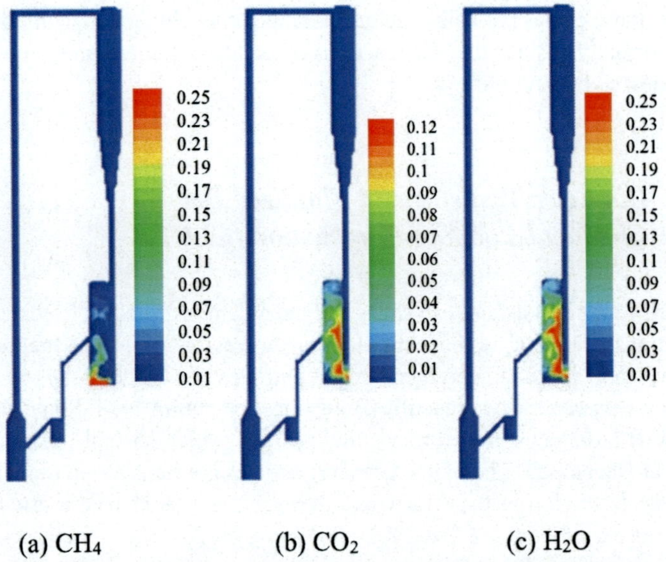


**Fig. 6.14** Instantaneous volume fraction of gas species $CH_4$, $CO_2$ and $H_2O$ ($t = 6.0$ s)

different reactions in the CLC system. In the FR, the reaction models can generally handle homogeneous gas phase reactions, gas phase reactions catalyzed by the solids and heterogeneous reactions involving the solids as reactants. Nth order kinetic formulations are applied for intrinsic kinetics and the heterogeneous reactions are dependent on the surface of reacting solids present in the control volume according to the shrinking core model.

$$CH_4(g) + NiO(s) \rightarrow Ni(s) + 2H_2(g) + CO(g) \quad \Delta H^0_{r1} = 203.75\,kJ/mol \tag{FR-R1}$$

$$H_2(g) + NiO(s) \rightarrow Ni(s) + H_2O(g) \quad \Delta H^0_{r2} = -2.1\,kJ/mol \tag{FR-R2}$$

$$CO(g) + NiO(s) \rightarrow Ni(s) + CO_2(g) \quad \Delta H^0_{r3} = -43.3\,kJ/mol \tag{FR-R3}$$

$$O_2(g) + 2Ni(s) \rightarrow 2NiO(s) \quad \Delta H^0_{r4} = -479.4\,kJ/mol \tag{FR-R4}$$

The reaction rates are listed in Table 6.7. The $CH_4$ reaction (Reaction FR-R1) embraces the direct oxidation of $CH_4$ with NiO and the subsequent oxidation of CO and $H_2$ gases. Reactions (FR-R2) and (FR-R3) are considered to determine the CO and $H_2$ concentration. Based on the kinetics rate model used in the present work covering the global oxidation of $CH_4$ to $CO_2$ and $H_2O$, the reactions (FR-R2), (FR-R3) and (FR-R4) proceed in the backward direction towards equilibrium.

In the AR the oxygen-carrier is regenerated by oxidation in air:

$$2Ni + O_2 \rightarrow 2NiO \quad \Delta H_{1200K} = -468 \quad kJ/mol\ O_2 \tag{AR-R1}$$

where $Y_{O2}$ represents the weight fraction of oxygen, and $M_{O2}$ is the molecular weight of oxygen.

The methane is used as fuel gas in the FR, air as fluidizing gas in the AR and metal oxide particles containing 18 wt. % NiO as oxygen carrier in a CLC system. The FR consisted of a bubbling fluidized bed with a bed diameter and height of 0.05 m and 0.1 m. The fluidizing gas consists of methane and nitrogen. The fluidizing gas in the LS was nitrogen. The diameter and height of the LS are 0.025 and 0.06 m. The AR consisted of a bubbling fluidized bed (0.05 m i.d.) with a bed height of 0.1 m, followed by a riser of 0.02 m i.d. and 1.0 m height. Secondary air was introduced at the top of the bubbling bed to help particle entrainment. A Ni-based oxygen carrier was used. The density of oxygen carrier was 2480 kg/m$^3$ with the diameter of 0.1–0.3 mm. Table 6.8 lists parameters used in present simulations.

The instantaneous solids volume fraction and volume fraction of $O_2$ gas species are shown in Fig. 6.13 in the CLC system. Fluidizing gas is introduced from the inlets of the AR, FR and loop seal. The solids volume fractions are high in the FR, LS and bubbling fluidized bed of the AR. The volume fractions of particles are low in the riser of the AR. The formation of bubbles is found in the bubbling fluidized

**Table 6.8** Parameter values used in simulations

| Item | Unit | Values | Item | Unit | Values |
|---|---|---|---|---|---|
| Particle diameter | mm | 0.2 | Particle density | $kg/m^3$ | 2470 |
| AR inlet primary gas flux | $m^3/h$ | 0.72 | FR inlet gas flux | $m^3/h$ | 0.17 |
| AR inlet secondary gas flux | $m^3/h$ | 0.15 | FR initial temperature | K | 1153 |
| AR initial temperature | K | 1223 | FR inlet gas species mass fraction | / | $CH_4$: 0.3, N2: 0.7 |
| AR inlet gas species mass fraction | / | $O_2$: 0.21, $N_2$: 0.79 | SL inlet gas species mass fraction | / | $N_2$: 1.0 |
| SL initial temperature | K | 1223 | SL inlet gas flux of SL | $m^3/h$ | 0.17 |

bed of the AR, while clusters are formed near the walls where the local solids volume fraction is large in the riser of the AR. Particles carried by gas flow up, hit the top wall and fall down in the riser with a high gas velocity. Simulated results show that the solids have the highest volume fraction near the walls and the lowest one in the center of riser. Particles enter the cyclone, and are separated from gas. Gas is discharged from the exit. Separated particles in the cyclone flow down along a dip leg by gravity, and transport into the FR. In the FR, small bubbles develop near the inlet and travel through the bed center regime, forming larger bubbles by coalescence. Finally, bubbles are broken at the bed surface. Gases flow out from the exit at the top of the FR. Particles having high volume fraction flow into the LS by gravity. In the LS, the velocity of nitrogen is low, and small bubbles are found. Particles move from the LS to the AR by gravity. Finally, the circulation of carrier material particles is established in the CLC system.

Air as fluidizing gas is fed from inlet at the bottom of the AR. The volume fraction of $O_2$ has the highest value in the bottom of bubbling fluidized bed and the lowest one at the top of the riser in the AR. This means a bubbling fluidized bed is necessary at the bottom to give a long enough residence time for particles to be oxidized to a high degree. Simulated results show that the volume fraction of $O_2$ has the highest value in the center and the lowest one near the walls where the volume fraction of particles is large. The oxidation reaction would reduce the volume by transferring oxygen from the gas phase to the solids phase. Thus, the volume fraction of $O_2$ decreases significantly at the bottom where enough solid material is available. The reduction of the oxygen carriers entering and leaving the AR is found. The difference in the degree of reduction between the inflow and outflow of the AR is increasing. This is due to the growing ability of the air reactor to regenerate the solid oxygen carrier at the largest degrees of reduction.

In the FR, the fuel gas, methane, reacts with the oxygen carrier to give $CO_2$, CO and $H_2O$ gas species. Figure 6.14 shows the instantaneous volume fraction of gas species of $CH_4$, $CO_2$ and $H_2O$ in the CLC system. The oxygen carrier was fluidized by fuel gas consisting of 30 wt.% $CH_4$ and 70 wt.% $N_2$ into the FR through a distributor. The upward flowing gas bubbles provide the energy to keep the oxide metal and fuel gas highly mixed. The reactant ($CH_4$) from the gas phase reacted with oxygen ($O_2$) of the metal oxygen carrier and was converted to gas products CO, $CO_2$ and $H_2O$. The bubble formation was captured by our CFD simulations. From simulations, we saw most of salient bubble features such as formation, rise and burst. These features influence the amount of fuel burned as fast bubbles lead to lower reactant conversion rate. The computed flow patterns due to the bubble predicted a global mixing between the gas phase and solids phase in the FR. The volume fraction of methane is decreased, while the volume fraction of $CO_2$ and $H_2O$ were increased along bed height in the FR. With an assumption of zero gas–solid contacting efficiency in the freeboard of the FR, the molar fractions of $CH_4$, $CO_2$ and $H_2O$ were constant values. Therefore, the gas by passing due to large bubbles was a cause of poor conversion rate in the FR.

### *6.2.2 Circulating Fluidized Bed Chemical-Looping Combustion*

Flow behavior of gas and oxygen carriers are simulated in a cold model of an interconnected fluidized bed for chemical-looping combustion system. The CLC system, seeing in Fig. 6.15a, has a circulating fluidized bed (AR) with an extra bubbling fluidized bed (FR) after the cyclone. The FR has a diameter of 0.052 m with a bed height of 0.5 m. The AR has a diameter of 0.02 m with a height of 2.0 m. The oxygen carrier has a diameter of 0.2 mm with a density of 3446 kg/m$^3$. The total solids inventory in the system was about 1.5 kg of solid material. Table 6.9 gives parameters used in experiment and present simulations. With oxidation in the AR, air with mass fraction 0.21 for oxygen gas is fed into the bed from the inlet distributors of the AR. While the inert nitrogen gas $N_2$ as fluidized gas is injected into the LS from the inlet distributor. With reduction in the FR, methane with mass fraction of 0.9 is fed into the distributor of the FR.

The contours plots of instantaneous volume fraction of particles are shown in Fig. 6.15b in the CLC system. At the beginning of the simulation, the FR, AR, and LS are partially pre-loaded with solids. Simulations show that particles in the AR are carried by air from the inlet to the top, and then enter into the cyclone. Since particles do not allow flow out from the exit of cyclone, the inventory of particles in the system is constant. Particles are separated in the cyclone, move down along a dip leg by gravity, and then transport into the FR. In the FR, particles are fluidized

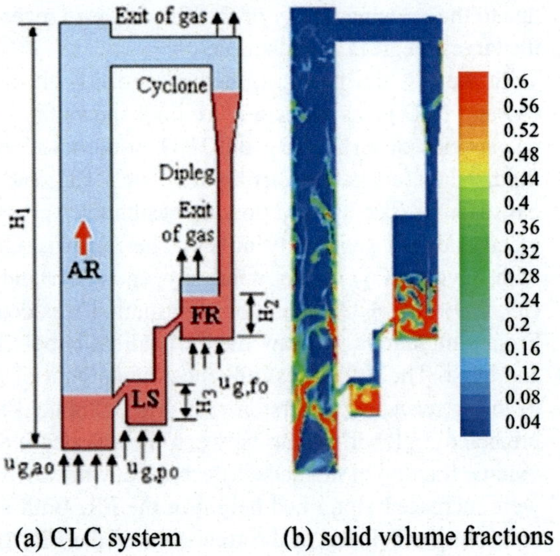


**Fig. 6.15** Instantaneous volume fractions of particles of the CLC system

**Table 6.9** Parameter used in simulations

| Description | Unit | Values | Description | Unit | Values |
|---|---|---|---|---|---|
| Width of AR, loop seal, FR | m | 0.19, 0.09, 0.13 | Height of AR, loop seal, FR | m | 3.0, 0.15, 0.5 |
| Initial solid height of AR, loop seal, FR | m | 0.45, 0.15, 0.15 | Initial solid volume fraction of AR, pot-seal, FR | | 0.5, 0.5, 0.5 |
| Initial temperature | K | 1173 | AR inlet gas velocity | m/s | 2.0 |
| AR inlet gas volume fraction | / | 1.0 | AR inlet gas temperature | K | 1173 |
| Inlet solid species mass fraction of AR | / | Ni:0.422, NiO: 0.578 | Inlet gas species mass fraction of AR | / | $O_2$: 0.23, $N_2$: 0.77 |
| Inlet gas velocity of FR | m/s | 0.12 | Inlet gas volume fraction of FR | / | 1.0 |
| Inlet solid species mass fraction | / | Ni: 0.422, NiO: 0.578 | Inlet gas species mass fraction of FR | / | $CH_4$: 0.9, $N_2$: 0.1 |

by gas, and then enter into the LS. In the LS, particles are transported to the AR by fluidizing inert nitrogen gas. Thus, the circulation of particles is formed in the CLC system.

Simulations show that the FR includes a dense region with high volume fraction of particles in the bottom and a dilute freeboard with low volume fraction of particles in the top of the bed. It can be seen there is an important change in the volume fraction profiles between the dense and the freeboard regions. The motion of bubbles was observed in the bed. These features influence the amount of fuel burned as fast bubbles lead to lower reactant conversion rate. The computed flow patterns due to the bubble predicted a global mixing between the gas phase and solids phase in the FR. The fluidizing gas provides the energy to keep the oxide metal and fuel gas highly mixed in the dense region.

In the AR, clusters are formed near the walls where the local volume fraction of particles is large. Particle clusters both in the center and near the walls are formed and dissolved dynamically. Since the solids acceleration rate is proportional to the inter-phase velocity difference, the particles in the center core region are accelerated quickly under high drag force due to high inter-phase velocity difference; however, the particles near the wall region are accelerated slowly under relatively low drag force due to low inter-phase velocity difference. Thus, the core-annulus structure is formed along radial direction. Simulations also show the solids volume fraction has the highest value in the bottom region, and the lowest one at the upper region of the AR. A dense region at the bottom and a dilute region at the top are formed along the AR height. Thus, the gas–solid system in the AR exhibits a more heterogeneous structure with a core-annulus structure along lateral direction, and dense in the bottom and dilute at the top along axial direction of the AR.

From the reduction reaction (FR-R1) in the FR, the reactant ($CH_4$) reacted with oxygen ($O_2$) of the metal oxygen carrier was converted to gas products $CO_2$ and $H_2O$. Figure 6.16 shows the contours plots of instantaneous weight fraction of gas species $CH_4$ and $O_2$ in the CLC system. The reduction has taken place in both the FR and the connecting pipe from the FR to the LS. In the FR, the reactant $CH_4$, fed from the inlet distributor at the bottom, reacts with oxygen carrier according to reaction (FR-R1). The same reaction has taken place at the LS. The weight fraction of methane is reduced along bed height in the FR. The simulated instantaneous weight fraction of gas specie $O_2$ has the highest value in the bottom of the AR since air is

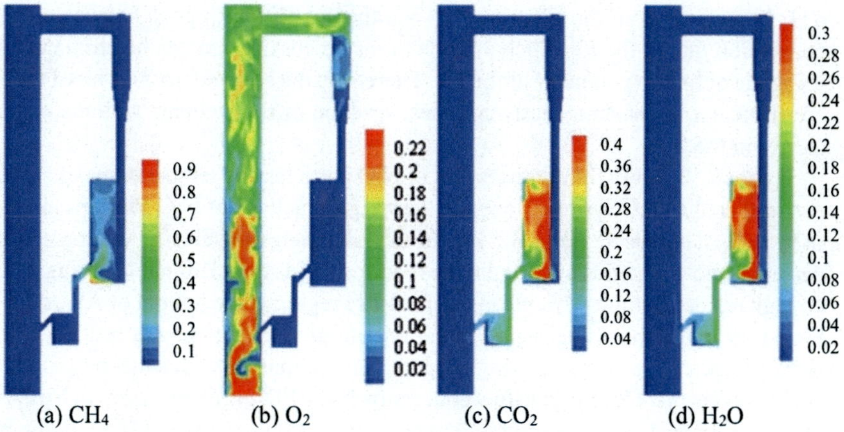


**Fig. 6.16** Instantaneous weight fraction of $CH_4$, $O_2$, $CO_2$ and $H_2O$ in the CLC system

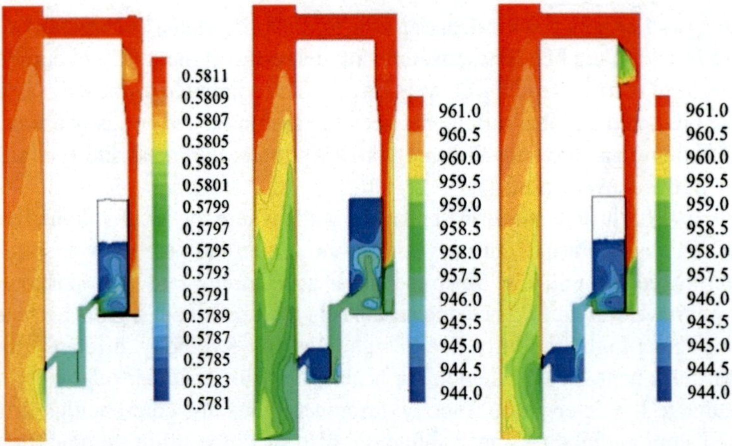


(a) NiO particles (b) gas temperature (c) particles temperature

**Fig. 6.17** Contour plot of weight fraction of NiO particles and temperatures in the CLC system

fed from the inlet distributor. The weight fraction of specie $O_2$ is decreased along the height of the AR due to reaction (AR-R1). We also found that the weight fraction of specie $O_2$ has the highest value in the center, and the lowest one near the walls because of the large reaction (AR-R1) where the solids volume fraction is high. The oxidation reaction continues in the cyclone due to the high temperature and solids volume fraction. These results indicate a stronger regeneration of the oxygen carrier passing through the air reactor.

The weight fractions of gas species $CO_2$ and $H_2O$ have the lowest value in the dense region, and the highest one in the freeboard of the FR. This means that the weight fraction of species $CH_4$ is reduced and the weight fractions of species $CO_2$ and $H_2O$ are increased along height in the dense region of the FR. The reactant $CH_4$ bypassing from the FR to the LS via the connecting pipe is a cause of poor conversion rate in the FR. Such a problem for process design can be eliminated by using the optimum geometry of the FR. Therefore, the selected geometries of the FR are sufficient to ensure a nearly complete reaction of the methane with the oxygen carrier in the FR.

Figure 6.17 shows the contours plot of NiO particles and temperatures of gas and particles in the CLC system. Roughly, the weight fractions of NiO particles increase in the AR. They decrease in the FR. In the AR, high gas velocity to carry out the necessary flow of solids gives them the required solids circulation rate during a long enough residence time. This means that a dense region at the bottom of AR could be necessary under the dilute region in the riser to give a long enough residence time for particles to be oxidized to a high degree. Simulations show that the low oxidation conversion in the AR limits the fuel conversion in the FR, even though enough oxygen in the solid material was transported to convert the fuel to $CO_2$ and $H_2O$. In the FR, simulations predicted that about half of the gas conversion in the dense bed occurs close to the gas distributor because of the highest rates of reaction and high gas–solid mass transfer in this region. However, the gas conversion is reduced in upper parts of the dense bed due to limitations in the gas transfer between bubble and emulsion. Thus, a relevant fraction of unconverted fuel bypasses the dense bed through bubbles.

### 6.2.3 *Dual Circulating Fluidized Bed Chemical-Looping Combustion*

Considering enough contact of fuel gas with oxygen carriers for the CLC process, both FR and AR are designed as bubbling bed and riser, respectively. The FR in form of bubbling fluidized beds makes fuel gas inside the bubble phase unconverted, which leads to the reduction of fuel combustion efficiency. For the dual circulating fluidized bed (DCFB) reactor system, both FR and AR use a circulating fluidized bed as the reactor (Shuai et al. 2012). The global circulation loop is closed by the lower loop seal and the overall circulation rate is dominated by the AR. The layout of the DCFB reactor system is displayed in Fig. 6.18a. By this way, we anticipate that there is a good gas–solid contact in the FR due to that the passing of the unconverted fuel gas inside the bubble is avoided, which has an enormous potential for scale up.

A Ni-based oxygen carrier here is selected due to the high reactivity and feasibility. The diameter and density of oxygen carrier are 135 μm and 3416 kg/m$^3$. The AR is 0.15 m in diameter and 4.1 m in height. The FR has a diameter of 0.159 m and a height of 3.0 m. The loop seal between FR and AR is used to prevent the gas mixing

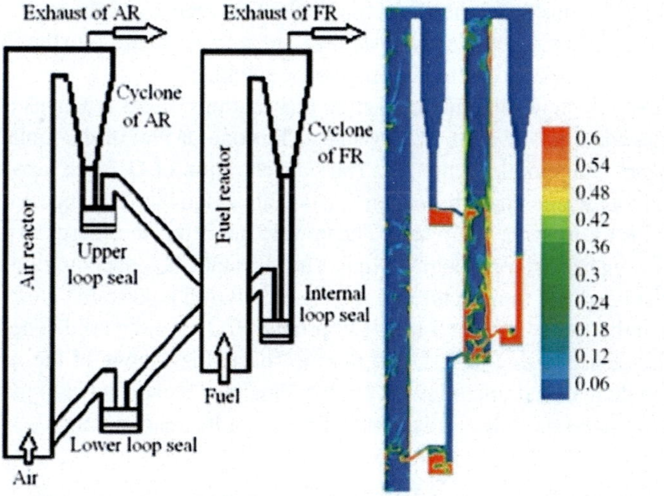


(a) Sketch of a DCFB (b) solids volume fraction

**Fig. 6.18** Instantaneous concentrations of particles in the DCFB system

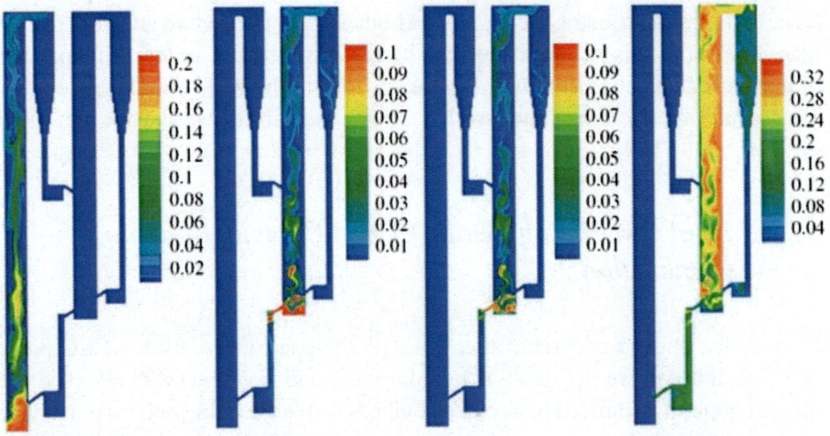


(a) $O_2$ (b) $CH_4$ (c) $H_2$ (d) $CO_2$

**Fig. 6.19** Instantaneous molar fractions of gas species

between the two reactors, which has a diameter of 0.15 m and a height of 0.25 m with the inlet gas velocity of 0.1 m/s. The diameter of cyclone of the AR and FR is 0.15 m. Both the down-comers have a diameter of 0.05 m.

Figure 6.18b displays the distribution of solid concentrations in the DCFB reactors. The non-uniform distribution of solid concentrations can be recognized in the AR and FR. It can be observed that the formation of the clusters exists near the walls of the reactors. For the center region, the concentration of particles displays a low level. Some oxygen carriers are driven into the cyclone by the gas. As it can be seen, there is an obvious difference between the bottom region and the top region, which limits the combustion efficiency. Overall, solid volume fractions in the FR are higher than that in the AR and more clusters can be found.

Figure 6.19 shows the molar fraction distributions of gas reactants and products under the condition of quasi-steady state. The contour plot of the molar fraction of $O_2$ appears a temporal fluctuation. The concentration of $O_2$ decreases evidently at the locations where enough oxygen carriers are available. The reactant $CH_4$ is fed into the FR through the distributor. The upward gas offers the energy to preserve fuel gas and oxygen carriers enough mixing. The reactant $CH_4$ concentration is decreased along the height of the reactor and there is nearly full conversion close to the wall. Accordingly, the concentration of gas product $CO_2$ is increased. The distribution of $H_2$ is similar to that of $CH_4$. At the top space, the change of the gas species is not as evident as that at the lowest height. This may be attributed to that the lowest concentration of particles leads to the decrease of the reaction rate.

### 6.2.4 Fluidized Bed Chemical Looping Reforming (CLR)

Hydrogen is regarded as the most promising clean energy and plays a key role in the reduction of greenhouse gas emissions. The production of hydrogen has attracted great attention. Among these methods of hydrogen production, methane steam reforming on solid catalysts is a well-known industrial route. However, the hydrogen production by means of methane steam reforming will lead to appreciable amounts of $CO_2$ released. Chemical looping reforming (CLR) utilizes oxygen carriers to transport oxygen instead of direct combustion of fuel and air, which avoids the contact of fuel gas and air (Shuyan et al. 2008b). The CLR advantage lies in that the product $CO_2$ can easily be recovered and the heat needed for hydrogen production can be supplied by the oxygen carrier flow from the exothermic air reactor (AR) to endothermic fuel reactor (FR), without costly oxygen combustion.

The CLR system showed in Fig. 6.20a consists of two fluidized beds. Both the AR and FR include a bubbling fluidized bed with a bed height of 0.1 m and a bed diameter of 0.05 m. The AR is connected to a riser with a bed height of 1 m and a bed diameter of 0.02 m. The loop seal is located between the AR and FR to avoid the leakage of fuel (Shuai et al. 2013, 2014c, 2016a). The main geometries and operating conditions of the FR and AR are summarized in Table 6.10.

The metallic Ni is used as catalyst in most commercial steam reforming processes because of high reactivity, and low attrition rate. The reducing reactions of Ni-based oxygen carriers with fuel gas in the FR are considered as follows:

$$CH_4(g) + NiO(s) \rightarrow Ni(s) + 2H_2(g) + CO(g) \qquad \Delta H_{r1} = 203.75\ \text{kJ/mol} \tag{R22}$$

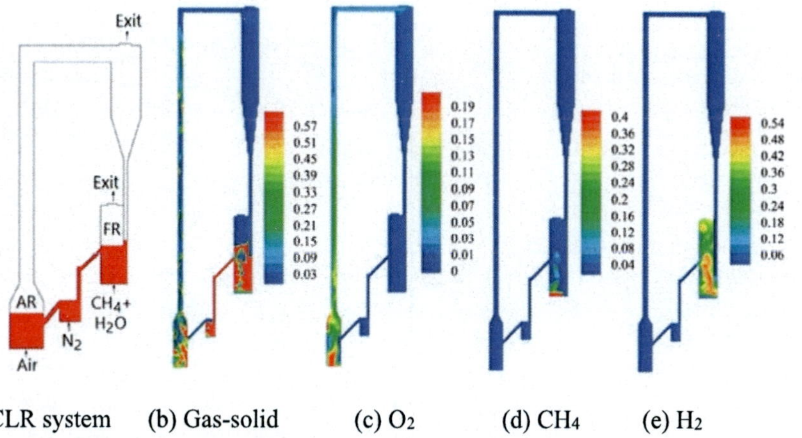


**Fig. 6.20** Instantaneous solid volume fraction and molar fraction of gas species at 10 s ($H_2O/CH_4$ = 0.3; $u_{FR}$ = 0.1 m/s; $T_{FR}$ = 1173 K; $T_{AR}$ = 1223 K)

**Table 6.10** Physical properties and model parameters used in the simulations

| Description | AR | FR | Description | AR | FR |
|---|---|---|---|---|---|
| Particle diameter (μm) | 200 | 200 | Particle density (kg/m$^3$) | 2500 | 2500 |
| Reactor height (m) | 0.15, 1 | 0.25 | Reactor diameter (m) | 0.05/0.02 | 0.052 |
| Initial concentration of particles | 0.5 | 0.5 | Initial temperature (K) | 1223 | 1073, 1123, 1173 |
| Inlet gas flow (m/s) | 0.46/0.59 | 0.07, 0.1, 0.15 | Inlet gas composition (molar ratio) | $O_2$:$N_2$ 0.21:0.79 | $CH_4$:$H_2O$:$N_2$ 0.5:0.05:0.45 |

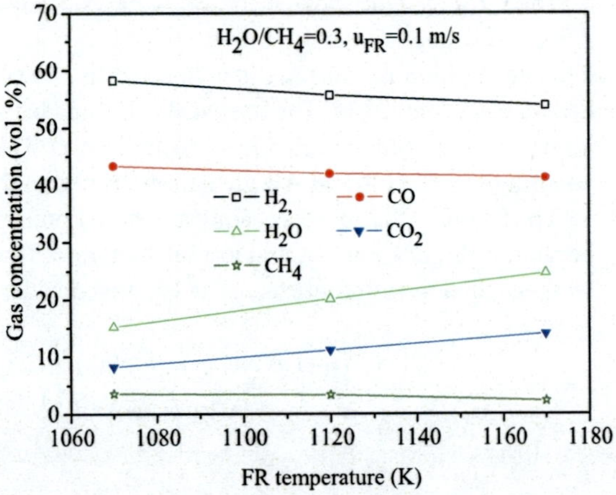


**Fig. 6.21** Effect of FR temperature on gas reactant and product concentrations at the outlet of FR

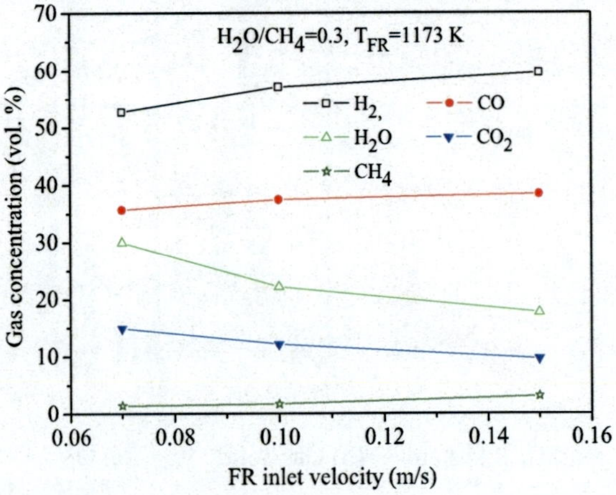


**Fig. 6.22** Effect of FR inlet velocities on product concentrations at the outlet of FR

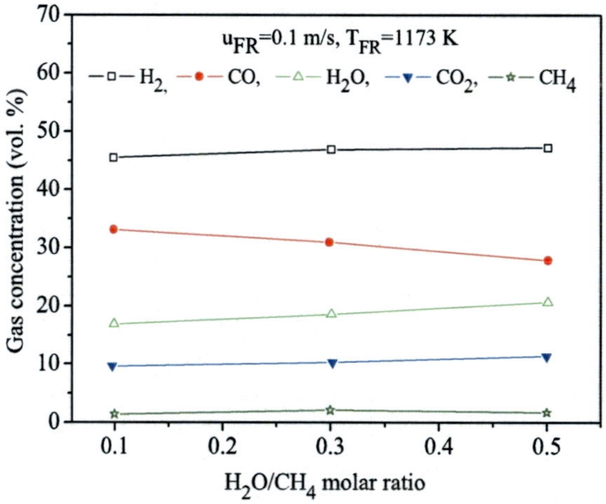


**Fig. 6.23** Effect of $H_2O/CH_4$ molar ratios on product concentrations at the outlet of FR

$$\begin{gathered} H_2(g) + NiO(s) \rightarrow Ni(s) + H_2O(g) \\ \Delta H_{r2} = -2.1\ kJ/mol \end{gathered} \tag{R23}$$

$$\begin{gathered} CO(g) + NiO(s) \rightarrow Ni(s) + CO_2(g) \\ \Delta H_{r3} = -43.3\ kJ/mol \end{gathered} \tag{R24}$$

The reduced particles are transferred to the AR and reoxidized:

$$\begin{gathered} O_2(g) + 2Ni(s) \rightarrow 2NiO(s) \\ \Delta H_{r4} = -479.4\ kJ/mol \end{gathered} \tag{R25}$$

The chemical reaction can be assumed as the main resistance to determine the global reaction rate. The shrinking-core model is adopted to determinate the reaction rates, which takes the following form:

$$-R_i = k_0 \exp\left(-\frac{E_0}{RT}\right) C_g^n \tag{6.13}$$

where $C_g$ and $n$ represent the concentration of the gas reactant and the reaction order. The detailed kinetic parameters are given in Table 6.11.

**Table 6.11** Kinetic parameters for Ni-based oxygen carriers

| | $CH_4$ | $H_2$ | CO | $O_2$ |
|---|---|---|---|---|
| $k_0$ ($mol^{1-n}\ m^{3n-2}\ s^{-1}$) | 0.2 | 0.15 | 0.059 | 0.84 |
| $E_0$ (kJ/mol) | 5 | 5 | 5 | 22 |
| $n$ | 0.2 | 0.4 | 0.6 | 0.7 |

During the methane steam reforming process, there will carbon formation due to the decompositions of hydrocarbons. The following three reactions are considered to represent methane steam reforming.

$$CH_4 + H_2O \rightarrow 3H_2 + CO \quad \Delta H_{r5} = 206.0\ \text{kJ/mol} \tag{R26}$$

$$CO + H_2O \rightarrow H_2 + CO_2 \quad \Delta H_{r6} = -41.0\ \text{kJ/mol} \tag{R27}$$

$$CH_4 + 2H_2O \rightarrow 4H_2 + CO_2 \quad \Delta H_{r7} = 164.9\ \text{kJ/mol} \tag{R28}$$

The kinetic model can be used to represent the methane steam reforming and water–gas shift reactions during the chemical looping reforming, and the corresponding reaction rate equations are expressed as

$$R_5 = k_5 \frac{\left(P_{CH_4}P_{H_2O}/P_{H_2}^{2.5}\right) - \left(P_{CO}P_{H_2}^{0.5}/K_5\right)}{DEN^2} \tag{6.14}$$

$$R_6 = k_6 \frac{\left(P_{CO}P_{H_2O}/P_{H_2}\right) - \left(P_{CO_2}/K_6\right)}{DEN^2} \tag{6.15}$$

$$R_7 = k_7 \frac{\left(P_{CH_4}P_{H_2O}^2/P_{H_2}^{3.5}\right) - \left(P_{CO_2}P_{H_2}^{0.5}/K_7\right)}{DEN^2} \tag{6.16}$$

$$DEN = 1 + K_{CH_4}P_{CH_4} + K_{H_2}P_{H_2} + K_{CO}P_{CO} + K_{H_2O}P_{H_2O}/P_{H_2} \tag{6.17}$$

where $P_i$ is the partial pressure of gas species $i$, $k_j$ and $K_j$ are the kinetic constant and the equilibrium constant of reaction $j$, respectively, and $K_i$ is the adsorption constant of gas species $i$. These constants follow an Arrhenius type dependence with temperature. The parameters are showed in Table 6.12.

$$k_j = k_{0j}\exp\left(-\frac{E_j}{RT}\right) \tag{6.18}$$

**Table 6.12** Parameters for steam reforming

| | | | |
|---|---|---|---|
| $k_{05}$, mol/(kg s) | $2.0 \times 10^{15}$ | $E_5$ (kJ/mol) | 275 |
| $k_{06}$, mol/(kg s) | $3.3 \times 10^{5}$ | $E_6$ (kJ/mol) | 86 |
| $k_{07}$, mol/(kg s) | $7.2 \times 10^{8}$ | $E_7$ (kJ/mol) | 144 |
| $K_{CO}$ (1/bar) | 40.91 | $K_{H2}$ (1/bar) | 0.0296 |
| $K_{CH4}$ (1/bar) | 0.1791 | $K_{H2O}$ (1/bar) | 0.4152 |
| $K_{05}$ (1/bar) | $1.17 \times 10^{13}$ | $\Delta H_5$ (kJ/mol) | 223,077.28 |
| $K_{06}$ (1/bar) | $1.77 \times 10^{-2}$ | $\Delta H_6$ (kJ/mol) | −36,583.68 |

$$K_i = K_{0i}\exp\left(-\frac{\Delta H_i}{RT}\right) \tag{6.19}$$

The solid oxygen carriers are loaded with an initial solid concentration of 0.5 in the AR and FR at the beginning. The restitution coefficient of particles and specularity coefficient are 0.97 and 0.5, respectively. It is assumed that all Ni in oxygen carriers are in the form of free NiO and the initial fraction of oxidation state is 0.1. The velocity-inlet conditions are set at the bottom of reactors and loop seal. At the top of separator, the pressure is set to be 1 atm. For the wall, no slip condition is applied and the wall temperature is considered as a constant.

The instantaneous contour plots of solid concentration and gas species molar fraction are shown in Fig. 6.20b–e at the quasi-equilibrium state. The bubble phase in the reactor can be clearly observed, which leads to the non-uniformity of oxygen carriers in the bed and has a great impact on the fuel conversion. The distribution of $O_2$ molar fraction in the AR reflects the regeneration process of oxygen carriers. With the bed height increased, the oxygen is continually transferred to oxygen carriers and shows a descending trend. Above the bed of the AR, there is a local increase in the $O_2$ molar fraction due to the inlet of the secondary air. For the FR, the highest $CH_4$ concentration value at the center region of bed, which is due to that the bubble passing results in gas leakage. At the top area, $CH_4$ concentration closes to zero, which means that nearly complete fuel conversion can be achieved. The $H_2$ molar fraction has a slight low value where the highest concentrations of oxygen carriers promote the conversion of $H_2$ to $H_2O$ near the wall.

The operating temperature is an important factor for the CLR process. To evaluate the effect of temperatures on the reaction, the FR temperature is changed from 1073 to 1173 K. From Fig. 6.21, it can be found that a high temperature promotes the $CH_4$ conversion and more $CO_2$ and $H_2O$ are captured. Accordingly, the $H_2$ yield is restrained. This may be attributed to the fact that the reactivity of oxygen carriers is enhanced and more fuel gas can be completely oxidized with increasing temperature.

To investigate the effect of operating velocity on the CLR performance, the FR inlet velocity is varied from 0.07 to 0.15 m/s. The gas reactant and product concentrations at the FR exit, as a function of the FR operating velocity, are shown in Fig. 6.22. An increase in the FR inlet velocity leads to a slight reduction of $H_2O$ and $CO_2$ concentrations, while there is a reverse influence on $H_2$ and CO concentrations. The fuel gas flow rate increases as the FR inlet velocity increases, which means that more fuel gas is available in the reactor and methane steam reforming reactions are enhanced. Hence, the yield of $H_2$ is increased. On the other hand, with increasing fuel flow, there are not enough oxygen carriers to react with fuel gas, which will lead to the reduction of fuel conversion and more $CH_4$ is obtained at the exit of FR.

The inlet gas composition of the FR includes of 50 vol % of $CH_4$ and 50 vol % of $H_2O + N_2$. To study the influence of the $H_2O/CH_4$ molar ratio, the molar fraction of the feeding $CH_4$ was maintained at a constant of 0.5 and the molar fraction of $H_2O$ is varied from 0 to 0.25. $N_2$ is used as a balanced gas. The effect of the $H_2O/CH_4$ molar ratio on the gas reactant and product concentrations at the FR exit is shown in Fig. 6.23. An increase in the $H_2O/CH_4$ molar ratio means more water vapor available for methane reforming reaction. It can be observed that the yield of $H_2$ is enhanced when the $H_2O/CH_4$ molar ratio is increased. Accordingly, the concentrations of CO and $CH_4$ are decreased and the molar fractions of $CO_2$ and $H_2O$ are increased. This may be attributed to that a high $H_2O/CH_4$ molar ratio promotes the water–gas shift reaction. However, compared with the previous two factors, $H_2O/CH_4$ molar ratio has less influence on the performance of FR.

## 6.3 Fluidized Bed $CO_2$ Capture

Carbon dioxide capture from combustion and utilization of carbon-based fuel, such as fossil fuel, plays a critical role in solving the urgent climate change problem created by $CO_2$ emission. In coal fired power plants, $CO_2$ capture is one of the key technologies required to stabilize atmospheric $CO_2$ concentrations. For post-combustion capture the $CO_2$ can be separated using a variety of technologies including solvent absorption, adsorption or membrane based processes. The metal-based sorbents including $K_2CO_3$ and $Na_2CO_3$ are sorbents to remove $CO_2$ because they are inexpensive, low energy demand, no corrosion to device, no secondary pollution and have high activity with $CO_2$.

The sorption looping cycles for $CO_2$ capture could involve a carbonation reactor and a regeneration reactor operating as chemical reactors. The carbonator fluidized bed receives a stream of flue gases and the $CO_2$ reacts with sorbents in an exothermic reaction. Sorbent regeneration takes place in the regenerator reactor which consists of a fluidized bed to produce a concentrated stream of $CO_2$ suitable for capture. Due to the strongly endothermic calcination reaction and the heating up of the incoming solids from the carbonator up to the required reaction temperature, the higher capture efficiency of carbon dioxide leads to an augmented sorbents feed rate into the carbonator, consequently the solid circulation within the system must be intensified.

The $CO_2$ gas sorption-regeneration circulating fluidized bed sorber (CFBS) system is shown in Fig. 6.24a. It consists of a riser as the sorber zone, a cyclone as a separator of solid sorbents, regenerator as the sorption zone and a loop seal for solids transport from the regenerator to the riser (Shuai et al. 2016b). This system has the sorption (or absorption) and regeneration (or desorption) reaction in a series connection form. It can handle large volumetric flow rate of flue gases. The flue gases are fed at the bottom of the riser which is designed to be a sorption zone. In this section, $CO_2$ and steam in the flue gases are absorbed into the solid sorbents. The remaining gases leave the system through the exit of the cyclone at the top of the regenerator. Two airlock rotary valves will be designed to place them between the separator and the regenerator and between the regenerator and the loop seal, respectively, to prevent the gas leakage in the regenerator. The absorbed solid sorbents are decomposed in the regenerator by reducing the pressure. The water vapor and the carbon dioxide, which can be easily separated by conventional cooling process, leave the system via the outlet port at the top of the regenerator by means of a vacuum

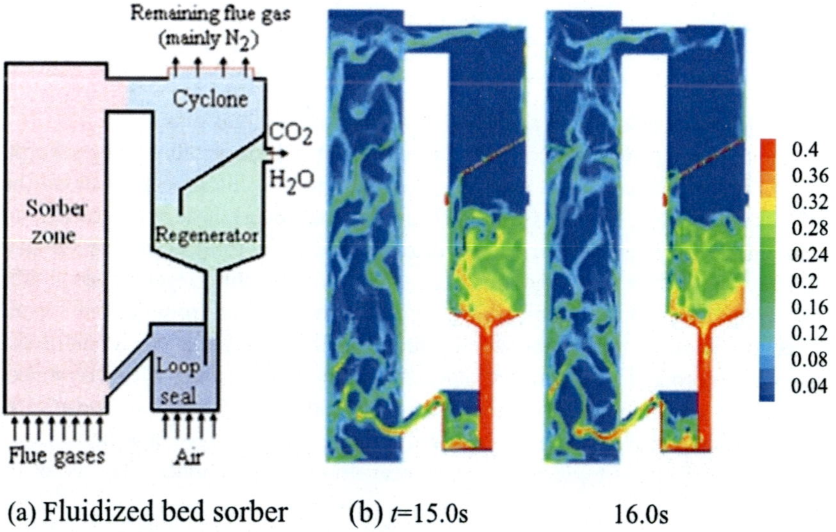


**Fig. 6.24** Instantaneous distributions of volume fractions of sorbent particles

fan. Then, fresh solid sorbents are sent back to the riser through a loop seal and again continue the sorption-regeneration cycle. At the bottom of the loop seal, small amount of air is fed at minimum fluidization velocity to assist the circulation of solid sorbents.

Flue gas consists of $CO_2$ and steam gases. Potassium carbonate ($K_2CO_3$) is chosen as the solid sorbent. For simplicity, the following hypotheses are considered: the solids phase is characterized by a mean particle diameter and density. Both phases are continuous assuming a single gas phase and a single solids phase. In this sorption zone of the riser, $CO_2$ and steam in the flue gases react with potassium carbonate particles, and are absorbed into the solid sorbents. The absorbed solid sorbents are regenerated in the regenerator, where the absorbed solid sorbents are decomposed by reducing the pressure. The following reaction occurs in the CFBS system, respectively:

$$K_2CO_3 + CO_2 + H_2O \rightarrow 2KHCO_3 + \text{heat} \tag{R29}$$

$$2KHCO_3 \rightarrow K_2CO_3 + CO_2 + H_2O - \text{heat} \tag{R30}$$

The data needed for modeling the absorption and desorption is the reaction rate equation. The sorption reaction rate (R29) is expressed as a first order reaction with respects to $CO_2$ concentration and volume fraction of $K_2CO_3$ solid sorbent.

$$R_1 = -k_1 C_{CO_2} \alpha_s \tag{6.20}$$

$$k_1 = 0.55\exp\left(-\frac{3609}{RT_g}\right) \tag{6.21}$$

where $R$ is the gas constant and $T_g$ is the temperature of gas phase.

Strategies like pressure swing adsorption (PSA) and temperature swing adsorption (TSA) processes have been used for adsorption in a cyclic process. Among them, PSA is a cyclic process of adsorption/desorption that occurs through pressure changes and can be very suitable for carbon dioxide separation from exhaust gases due to its easy application in a large pressure range. In the sorption-regeneration fluidized bed system, the potassium bicarbonate particles transported from cyclone separator is decomposed in the regenerator by reducing the pressure at the vacuum swing adsorption processes. The water vapor and the carbon dioxide will be removed by means of a vacuum fan. Depending on the temperature, partial pressure and adsorbent pore sizes of potassium bicarbonate solid sorbent, and $CO_2$ gas can be adsorbed. The mass transfer rate is represented by a linear driving force model. The rate of mass transfer is

$$R_2 = K_2(Y^*_{KHCO_3} - Y_{KHCO_3}) \tag{6.22}$$

$$K_2 = 0.01\varepsilon_s \tag{6.23}$$

where $K_2$ is the overall mass transfer coefficient and $Y^*$ is the maximum adsorbed potassium bicarbonate species concentration at equilibrium given by the adsorption isotherm. The overall mass transfer coefficient is assumed to be a function of solids volume fraction $\varepsilon_s$ in the regenerator.

The main parameters of operating conditions of the FR and AR are summarized in Table 6.13. The initial static bed heights in the riser, regenerator and loop seal are constant. At the inlet, all velocities and volume fractions of gas and solids phases are

**Table 6.13** Parameters used for numerical simulations

| Description | Value | Description | Value |
|---|---|---|---|
| Height of sorber (m) | 3.0 | Diameter of sorber (m) | 0.6 |
| Particle diameter (μm) | 210 | Particle density (kg/m$^3$) | 1530 |
| Gas density (kg/m$^3$) | 1.2 | Gas viscosity (kg/ms) | $1.789 \times 10^{-5}$ |
| Diffusion coefficient (cm$^2$/s) | $O_2 = 0.178$, $CO_2 = 0.1381$, $H_2O = 0.2178$, $N_2 = 0.2$ | Inlet mass fraction of gas species in sorber | $O_2 = 0.161$, $CO_2 = 0.225$, $H_2O = 0.092$, $N_2 = 0.522$ |
| Inlet gas velocity of sorber/loop seal (m/s) | 1.0/0.1 | Inlet gas temperature of sorber/loop seal (K) | 343.15/298.15 |
| Inlet mass fraction of gas species in loop seal | $O_2 = 0.23$, $N_2 = 0.77$ | Initial mass fraction of sorbents | $K_2CO_3 = 1$, $KHCO_3 = 0.0$ |

specified in the riser and loop seal. The outlets of the cyclone and the regenerator are modeled using a continuity condition for all variables, except for gas pressure.

Figure 6.24b shows the instantaneous solid sorbent volume fraction at the superficial gas velocities of 1.0 and 0.1 m/s in the sorber zone and loop seal of the sorption-regeneration fluidized bed system. The solids volume fraction has the lowest value in the sorber zone, and the highest one in the connecting pipe between the regenerator and the loop seal because of low gas superficial velocity. In the sorber zone, the formation and breakage of clusters which have high local solids volume fraction are found in the center region and near the walls. These clusters are dynamic. Simulations also indicate that the velocity of particles is negative near the walls. This means clusters near the walls flow down. Roughly, the volume fraction of sorbent particles has the highest value in the bottom, and the lowest one at the top of the sorber zone. This means that the solid volume fraction decreases along height. Sorbent particles are separated in the cyclone, and gas is discharged from the exit. At the top of the connecting pipe between the regenerator and loop seal, sorbent particles accumulate and fall down by gravity. At the bottom of the connecting pipe, particles are transported into the loop seal by means of fluidizing gas. The flow of bubbles is found by fluidizing gas from the inlet distributor in the loop seal.

Figure 6.25 displays the transient distributions of weight fraction of potassium bicarbonate species $KHCO_3$ at the superficial gas velocities of 1.0 and 0.1 m/s in the sorber zone and loop seal of the sorption-regeneration fluidized bed system. The flue gases are fed at the bottom of the sorber zone and sorbent particles are transported from the loop seal by a connecting pipe. In the sorber zone, $CO_2$ and steam in the flue gases are absorbed into the solid sorbents. The $KHCO_3$ weight fraction has the

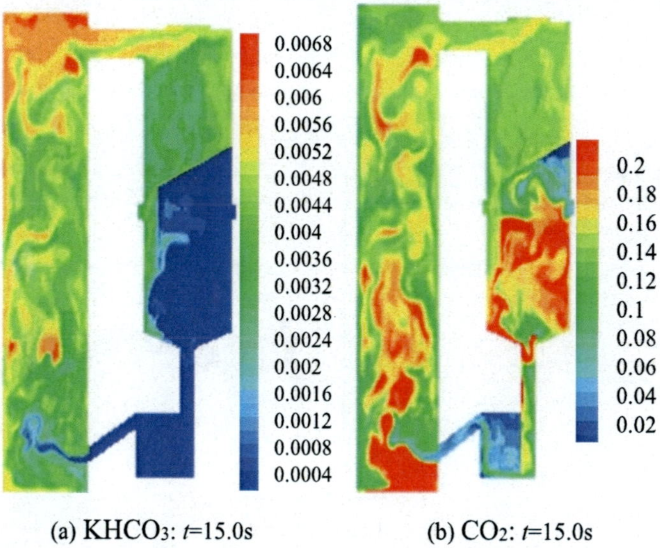


**Fig. 6.25** Instantaneous distributions of $KHCO_3$ and $CO_2$ species weight fraction

lowest amount in the bottom, and the highest one at the top. Clearly, the $KHCO_3$ weight fraction increases along riser height. The remaining gases leave the cyclone through the exit. The absorbed sorbent particles are regenerated in the regenerator. The $KHCO_3$ weight fraction is reduced. Exhaust $CO_2$ and steam leave the system via the outlet port at the top of the regeneration. Then, fresh sorbent particles are sent back to the riser through a loop seal, and continue the sorption-regeneration cycle. The $KHCO_3$ species weight fraction at the wall is higher because the sorbent particles are flowing downward, increasing the carbonation reaction of $K_2CO_3$ in the sorber zone.

In this system, $CO_2$ and steam are fed at the bottom of the riser, and are absorbed into the solid sorbents based on absorption reaction (R29). The $CO_2$ weight fraction has the highest amount in the bottom, and the lowest one at the top of the riser. This indicates that the $CO_2$ weight fraction decreases along riser height. The remaining $CO_2$ gases will be discharged from the exit of cyclone. Based on this stoichiometry, the theoretical or maximum weight percent of $CO_2$ on the solid sorbent can be up to 0.22. Roughly, simulations show that the $CO_2$ weight fraction has the highest value in the center and the lowest one near the walls in the riser. This flow behavior has a disadvantage because it will make the rate of $CO_2$ removal to be heterogeneous in the radial direction. In the regenerator, the absorbed solid sorbents are regenerated based on desorption reaction (R30). The $CO_2$ gas is released from $KHCO_3$ by means of reducing pressure. In the vacuum swing adsorption processes of the regenerator, the adsorbed $CO_2$ is recovered at sub-atmospheric pressures. Such vacuum recovery of adsorbed $CO_2$ is conducted in the regenerator. The high $CO_2$ weight fraction is found in the regenerator. Exhaust $CO_2$ and steam leave the system via the outlet port at the top of the regeneration. This adds to the motivation for the further development of adsorbents designed for processes with contribution to the separation of $CO_2$ from flue gases.

Figure 6.26 shows the instantaneous gas species $CO_2$ weight fraction at the outlet of the separator and regenerator at the superficial gas velocities of 1.0 for the riser and

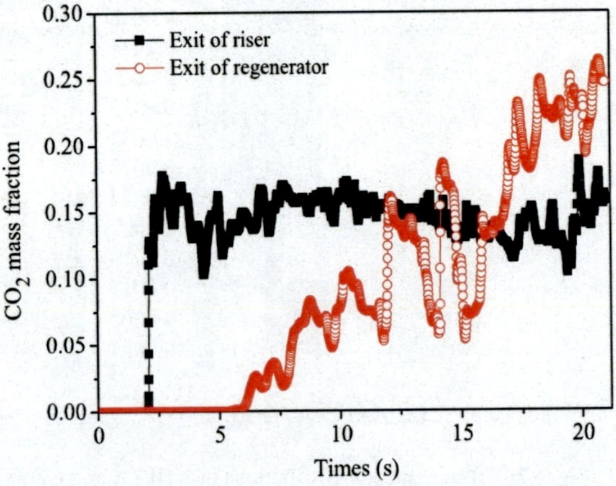


**Fig. 6.26** Instantaneous $CO_2$ weight fractions at the outlets of separator and regenerator

0.1 m/s for the loop seal. Cooled $CO_2$ gas species in flue gases enter the riser where the reaction of water vapor and carbon dioxide takes place. Thus, the volume fraction of $CO_2$ species increases toward the outlet of the riser. The weight fraction of $CO_2$ gas species at the outlet of the separator is oscillations due to the large variation of solids volume fraction by the formation and breaking of clusters in the sorber zone. Simulations show that the weight fraction of $CO_2$ gas species at the outlet of the regenerator oscillates since it depends upon solids mass flux from the separator. At the regenerator startup, the $CO_2$ mass fraction increases. The $CO_2$ mass fraction at the outlet of the regenerator is larger than that outlet of the separator. From simulated $CO_2$ fractions, the $CO_2$ recovery $R_{CO2}$ is calculated with the following expressions:

$$R_{CO2} = \frac{\text{moles of } CO_2 \text{ in product}}{\text{moles of } CO_2 \text{ in feed}} \tag{6.24}$$

The calculated $R_{CO2}$ is 43.8%.

Figure 6.27a shows a circulating fluidized bed $CO_2$ capture system (Khongprom and Gidaspow 2010). Cooled flue gases enter the sorption zone where the reaction of water vapor and carbon dioxide takes place and heat is liberated. Due to the equilibrium limitation, the sorbent has to be cooled to absorb the low temperature of $CO_2$ gas species. The exothermic reaction heat in the sorber zone can be used to regenerate the sorbent particles in the regenerator. The parameters used in the numerical simulations are listed in Table 6.14.

Figure 6.27b displays the transient distributions of $K_2CO_3$ solid sorbent volume fraction inside the circulating fluidized bed sorber system. The system is made up of the dense region near the bottom and the dilute region at the upper part of the sorber zone. The $CO_2$ and steam species in the flue gases are absorbed into the solid sorbents.

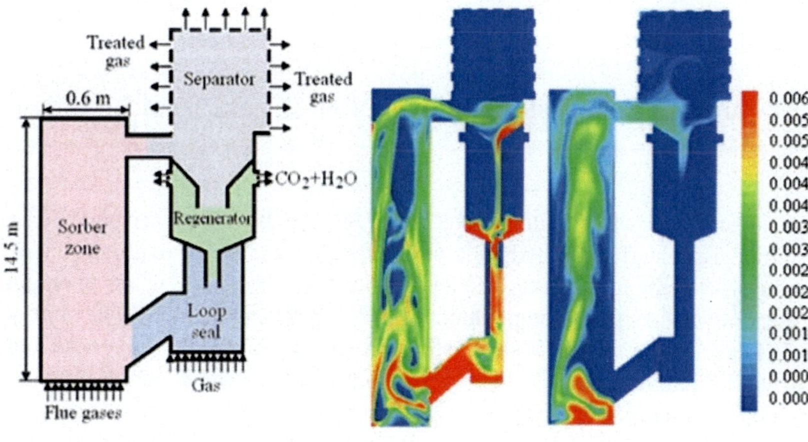


(a) Fluidized bed sorber (b) $t$=21.0s (c) averaged $CO_2$ concentration

**Fig. 6.27** Transient distributions of $K_2CO_3$ solid sorbent volume fractions

**Table 6.14** Parameters used for numerical simulations

| Description | Value | Description | Value |
|---|---|---|---|
| Particle diameter (μm) | 210 | Particle density ($kg/m^3$) | 1530 |
| Diffusion coefficient ($cm^2/s$) | $O_2 = 0.178$, $CO_2 = 0.1381$, $H_2O = 0.2178$, $N_2 = 0.2$ | Inlet mass fraction of gas species in sorber | $O_2 = 0.161$, $CO_2 = 0.225$, $H_2O = 0.092$, $N_2 = 0.522$ |
| Inlet gas velocity of sorber/loop seal (m/s) | 3.0/0.05 | Inlet gas temperature of sorber/loop seal (K) | 343.15/298.15 |
| Inlet mass fraction of gas species in loop seal | $O_2 = 0.23$, $N_2 = 0.77$ | Initial mass fraction of sorbents | $K_2CO_3 = 1$, $KHCO_3 = 0.0$ |

The $K_2CO_3$ solid sorbent enter the separator, and accumulates in the separator. The remaining gases leave the exit through exits the grid at the top of the separator. The separated sorbent particles fall into the desorption zone of the regenerator. Exhaust $CO_2$ and steam leave the system via the outlet port at the top of the regenerator. Then, fresh solid sorbents are sent back to the riser bottom through the loop seal and again continue the sorption-regeneration cycle. At the bottom of the loop seal, small amount of air at minimum fluidization velocity is fed to assist the circulation of solid sorbents.

The time-averaged contour of $CO_2$ concentration shows that the $CO_2$ concentration is high near the bottom of the sorber zone since the $CO_2$ sorption reaction rate depends on both the $CO_2$ concentration and the $K_2CO_3$ solid sorbent volume fraction. The dilution of $K_2CO_3$ solid sorbent volume fraction causes the slow decrease of $CO_2$ concentration. Conversely, the $CO_2$ concentration decreases drastically at the right wall at the exit port of the loop seal. The $CO_2$ concentration is very low with reduced pressure in the regenerator.

## 6.4 Supercritical Water Fluidized Bed (SCWFB) of Binary Mixture

The supercritical water fluidized bed (SCWFB) has been used for gasifying coal and biomass materials for hydrogen production, increasing the hydrogen yield and improve gasification efficiency. The supercritical water is a single, homogenous phase present above the critical temperature ($T_{CP} = 647$ K) and the critical pressure ($P_{CP} = 221$ bar). In the van der Waals theory, the critical point is the end of the liquid–gas equilibrium (coexistence) curve, the point in which liquid and gas merge into each other in a continuous manner and reach the same density, the critical density ($\rho_{CP} = 322$ $kg/m^3$). Near the critical point, the physical properties of water present drastic changes at temperatures from 600 to 700 K and pressures from 220 to 300 bar. For instance, the density decreases from the values for compressible liquid (about 750 $kg/m^3$) to values for dense gas (about 150 $kg/m^3$). These density variations near the critical point correlate with other macroscopic properties (molecular diffusivity, viscosity, and dielectric constant) and reflect changes at molecular level, such as molecular association by hydrogen bonding. Carrying out processes in compressible liquid water or dense vapor may result in energy savings and efficient separations. The supercritical water has unique properties, reflecting both the liquid and gas behavior, which makes it useful for industrial applications, such as thermochemical conversion of biomass for biofuel production, wastewater treatment, and hydrothermal synthesis of nanoparticles.

In coal gasification SCWFB, the bed materials are composed of coal particles of different sizes, in which segregation by size may occur during fluidization. Large particles (jetsam) component (JC) segregate at the bottom and small particles (flotsam) component (FC) float at the bed surface due to the collisional interaction between particles and hydrodynamic interaction between the fluid phase and the solids phase. Hence, an enhanced understanding of mixing/segregation of particles binary mixture will allow for an improvement of operations and design of SCWFB.

The two-dimensional SCW coal particles fluidized bed has width and height of 0.5 and 2.0 m along the axial direction z. The SCW is fed from the bottom distributor. The bed materials are composed of Geldart group B small coal particles component (FC) and Geldart group D large coal particles component (JC), respectively. The diameters of FC and JC are 0.65 and 1.0 mm with the same particles density of 2200 kg/m$^3$. Both the FC and JC are packed uniformly, and the initial bed height is 0.5 m from the distributor. The initial volume fractions of FC and JC are 0.25 and 0.25. The minimum fluidization velocities of the FC ($u_{mf,s}$) and JC ($u_{mf,c}$) are calculated from an empirical correlation (Wen and Yu 1966) according to the density and viscosity of SCW. The minimum fluidization velocity of the FC and JC binary mixture is (Huilin 2017):

$$u_{mf} = x_s u_{mf,s} + x_c u_{mf,c} \tag{6.25}$$

where $x_s$ and $x_c$ are the mass fractions of the FC and JC in the mixture. The values of minimum fluidization velocity of binary mixture are listed in Table 6.15.

The instantaneous volume fractions of SCW, FC and JC are shown in Fig. 6.28

**Table 6.15** Minimum fluidization velocities of FC ($u_{mf,s}$), JC ($u_{mf,c}$) and mixture ($u_{mf}$)

| Fluid | $T$(K) | $p$ (MPa) | $u_{mf,s}$ (cm/s) | $u_{mf,c}$ (cm/s) | $u_{mf}$ (cm/s) |
|---|---|---|---|---|---|
| SCW | 673 | 23.0 | 5.36 | 7.19 | 6.17 |
| | 673 | 25.0 | 4.80 | 6.41 | 5.51 |
| | 673 | 27.0 | 4.13 | 5.52 | 4.75 |
| | 673 | 29.0 | 3.36 | 4.51 | 3.87 |
| | 723 | 23.0 | 6.16 | 8.41 | 7.16 |
| | 773 | 23.0 | 6.54 | 9.06 | 7.65 |
| | 823 | 23.0 | 6.77 | 9.52 | 7.97 |

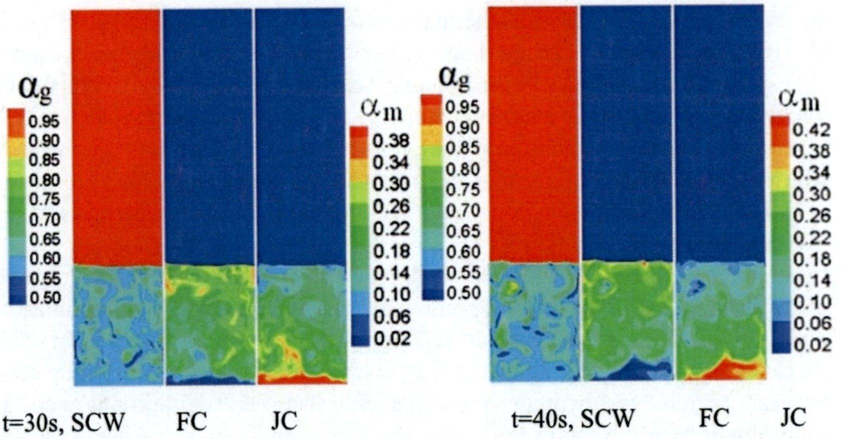


**Fig. 6.28** Instantaneous volume fractions of SCW, FC and JC at fluidizing velocity of $2.0u_{mf}$

at the fluidizing velocity $2.0u_{mf}$ with the pressure and temperature of 23.0 MPa and 673.0 K. Initially, both the FC and JC are packed uniformly at the initial bed height of 0.5 m from the bottom distributor. The initial volume fractions of FC and JC are 0.25 and 0.25. The mass fractions of FC and JC are 0.5 and 0.5 in the bed. The change with time of FC and JC volume fraction in the bed top and bottom layers is observed from numerical simulations along bed height. Both the largest FC volume fractions at the bed top and the highest JC volume fractions at the bed bottom are significant differences depending on the particle's properties of the binary mixture. It is also found that the small size bubbles form and pass through the bed. This vigorous movement of binary mixture with small bubbles contributes to the mixing of the FC and JC along the bed height.

From simulated instantaneous volume fractions, the time-averaged values are calculated. Figure 6.29 shows the distributions of time-averaged SCW, FC and JC volume fractions along bed height. Simulated results show that the FC volume fractions are low and the JC volume fractions are high near the bottom. Conversely, the JC volume fraction is low and FC volume fractions are high near the top. Three segments are identified from straight lines A, B and C according to the distribution of JC volume fractions. Four regions are determined from the intersections D and F from two straight lines. The segregation (SE) region exists with a high JC volume fractions and low FC volume fractions near the bottom. The JC volume fraction decreases and the FC volume fraction increases in the SE region. On the other hand, the FC and JC volume fractions are close to zero, and a freeboard (FR) region exists at the top the bed. The dense (DE) region is identified from the intersection D to F at the distance higher than SE region. The JC volume fraction is decreased and the FC volume fraction is increased in the DE region. The transition (TR) region is formed between the DE region and the FR region. Both FC volume fraction and JC volume fraction are decreased in the TR region. Numerical simulations further illustrate that

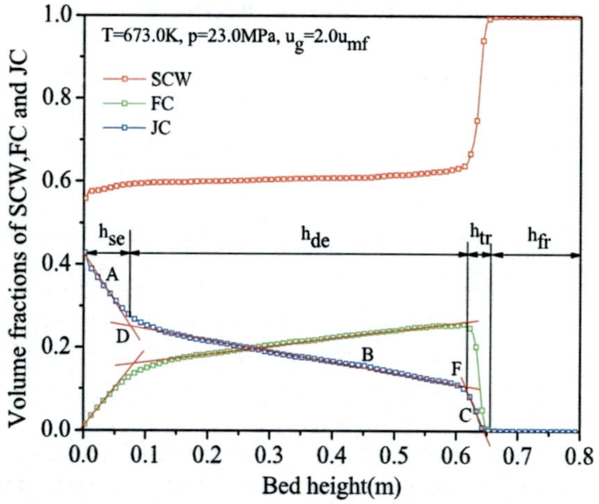


**Fig. 6.29** Distributions of SCW, FC and JC volume fractions along bed height

SCW volume fractions increase in the SE region, and keeps constant along the DE region height.

The distribution of mixing/segregation is determined using an integrated quantity of segregation index $s$ according to the mass fractions $x_{fc}$ of FC and $x_{jc}$ of JC in the bed (Wu and Baeyens 1998; Rao et al. 2011).

$$s = \frac{< S_{fc} > / < S_{jc} > -1}{(2 - x_{fc})/(1 - x_{fc}) - 1} \tag{6.26}$$

$$< S_{fc} >= \frac{\sum_{k}^{N_{cell}} (x_{fc}\alpha_{fc,k} h_k V_k)}{\sum_{k}^{N_{cell}} (x_{fc}\alpha_{fc,k} V_k)} \quad \text{and} \quad < S_{jc} >= \frac{\sum_{k}^{N_{cell}} (x_{jc}\alpha_{jc,k} h_k V_k)}{\sum_{k}^{N_{cell}} (x_{jc}\alpha_{jc,k} V_k)} \tag{6.27}$$

where $h_k$ represents the height of the centre of cell $k$ from the distributor. $V_k$ is the cell volume. The value of segregation index $s$ is 1.0 for a perfect segregation. While $s = 0$ represents a perfect mixing of FC and JC in the bed. Hence, the segregation index is used to quantify the degree of mixing/segregation of JC and FC in SCWFBs.

From the slope of the straight line of the curve of the FC volume fractions $\alpha_{fc}$ and JC volume fractions $\alpha_{jc}$ in the DE region along the bed height, the synergy angle defines as follows:

$$\eta_m = \left| arctg(\frac{d\alpha_{fc}}{dh}) - arctg(\frac{d\alpha_{jc}}{dh}) \right| \tag{6.28}$$

The information entropy is used to describe statistical phenomena in terms of uncertainty about the outcome. The information entropy assesses the degree of mixing of mixture. The greater information entropy corresponds to the better mixing degree of mixture. Thus, the information entropy appears to be a useful parameter for describing mixing efficiency in a system. The local information entropy of binary mixture of particles is calculated in each grid $k$:

$$s_k = -(x_{sc,k} \log x_{sc,k} + x_{lc,k} \log x_{lc,k}) \tag{6.29}$$

where $x_{sc}$ and $x_{lc}$ are the fractions of SC and LC in the cell. The mixture information entropy at the bed height $h$ is:

$$S(h) = -\sum_k (x_{sc,k} \log x_{sc,k} + x_{lc,k} \log x_{lc,k}) = S_{sc}(\alpha_{sc}) + S_{lc}(\alpha_{lc}) \tag{6.30}$$

where $S_{sc}$ and $S_{lc}$ are the SC and JC entropy parts, and they are computed from JC and SC volume fractions at the height $h$. The rate of the mixture information entropy along height $h$ is

$$\frac{\partial S(h)}{\partial h} = \frac{\partial S_{sc,h}(\alpha_{sc})}{\partial h} + \frac{\partial S_{lc,h}(\alpha_{lc})}{\partial h} \tag{6.31}$$

Comparing to Eq. (6.29), we see that the synergy angle may represent the rate of mixture information entropy $S(h)$ in the DE region. It enables evaluation of the degree of mixing/segregation based on the JC and SC volume fractions. Hence, it is observed that the simulated synergy angle shows a trend similar to that for the segregation index in the DE region.

The perfect mixing and segregation patterns of FC and JC binary mixture are shown in Fig. 6.30 along bed height. The differential quantity of synergy angle represents the slope of FC and JC volume fractions in the DE region. The value of synergy angle of 0.0° represents perfect mixing of FC and JC binary mixture. While the $\eta_m = 180°$ represents perfect segregation of FC and JC binary mixture. Thus, the

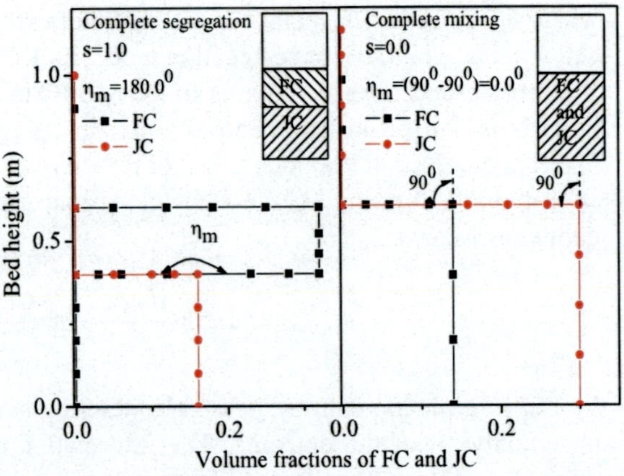


**Fig. 6.30** Segregation index and synergy angle of complete mixing and segregation of binary mixture

synergy angle measures the tendency of mixing/segregation of FC and JC along DE region height.

The instantaneous synergy angle is shown in Fig. 6.31 at the fluidizing velocity $2.0u_{mf}$ with the pressure and temperature of 23.0 MPa and 673.0 K. The synergy angle is small at the begin because of the perfect mixing of FC and JC binary mixture at the initial packing condition. The value of synergy angle increases when the numerical simulation runs to approximately 30 s of transient time. By this time, a pseudo-steady state is achieved. From simulated volume fractions of JC and FC, the averaged synergy angle is 25.12° at the fluidizing velocity $2.0u_{mf}$ with the pressure and temperature of 23.0 MPa and 673.0 K in the SCWFB.

The distributions of time-averaged SCW, FC and JC volume fractions are shown in Fig. 6.32 along bed height at the fluid pressures of 27.0 MPa and 29.0 MPa. Both

**Fig. 6.31** Profile of time-averaged synergy angles of FC and LC binary mixture

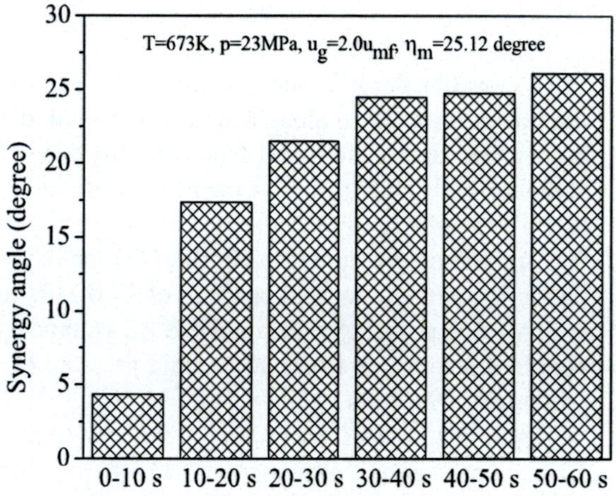


**Fig. 6.32** Profile of SCW, FC and JC volume fractions at two fluid pressures

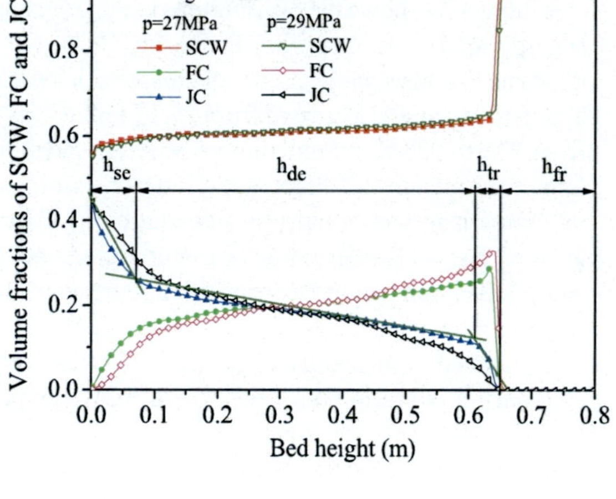

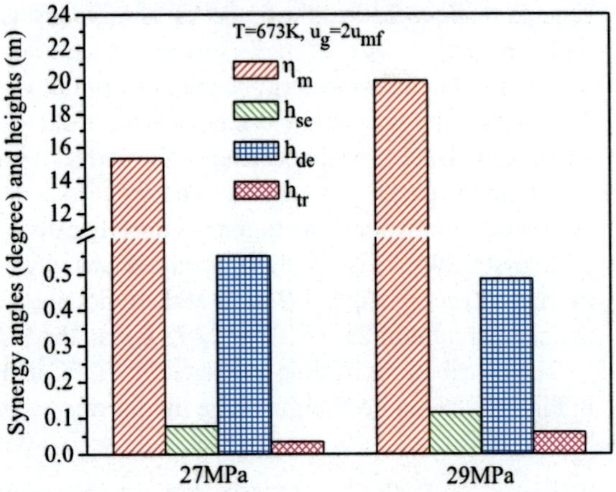


**Fig. 6.33** Profile of synergy angle and heights of SE, TR and DE at two SCW pressures

cases show that the JC volume fractions are decreased and the SCW and FC volume fractions are increased along bed height. The fluidization regime of binary mixture is characterized by the co-existence of a SE region near the bed bottom, FR region at the top bed, TR region near the bed surface and DE region within the mid of the bed. The height $h_{se}$ of the SE region are lower at the fluid pressure of 27.0 MPa than that at the fluid pressure of 29.0 MPa. On the other hand, the height $h_{de}$ of the DE region is larger at the fluid pressure of 27.0 MPa than that at the fluid pressure of 29.0 MPa. Comparing with simulated FC volume fractions in the SE region, the JC volume fractions are lower at the fluid pressure of 27.0 MPa than that at the fluid pressure of 29.0 MPa.

Figure 6.33 shows the profile of synergy angle and heights of the SE, TR and DE regions at two SCW pressures. The synergy angle, SE and TR heights increase with increasing SCW pressure. The DE height is reduced at the SCW pressure of 29.0 MPa. Simulated synergy angle is low, indicating that the mixing of binary mixture is improved at the SCW pressure of 27.0 MPa.

Figure 6.34 shows the distributions of synergy angle at three SCW pressures. The synergy angles increase with increasing SCW pressure. Simulated results further illustrate that the synergy angles decrease with increasing inlet fluidizing velocity $u_g$. The synergy angle decreases from 25.12 to 8.63° as the SCW temperature increases from 773 to 823 K at fluidizing velocity and pressure of 2.0$u_{mf}$ and 25.0 MPa. The integrated quantity of segregation index measures the degree of spatial separation in comparison with a uniform distribution of FC and JC. On the other hands, the synergy angle takes the volume fraction gradients of both FC and JC components along DE region height into account. It represents the variation of FC and JC volume fraction gradients.

Simulated results indicate that the SCW pressure has a strong effect on the mixing of binary mixture in comparison with SCW temperature effect. This seems to be

**Fig. 6.34** Distributions of synergy angle at three SCW pressures

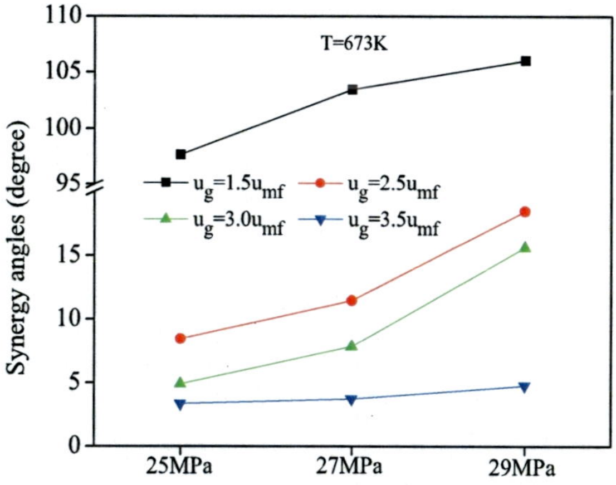


caused by the fact that the SCW density and viscosity varies with the change of pressure and temperature analyzed in present numerical simulations. Hence, more experimental measurements and numerical simulations are required to predict mixing/segregation of binary mixture with the change of SCW pressure and temperature in SCWFBs.

## References

Bougamra, A., & Huilin, L. (2014). Modeling of chemical looping combustion of methane using a Ni-based oxygen carrier. *Energy and Fuels, 28,* 3420–3429.

Guodong, L., Dan, S., Huilin, L., Bouillard, J., Yinghua, B., & Shuai, W. (2010). Computations of fluid dynamics of a 50 MWe circulating fluidized bed combustor. *Industrial and Engineering Chemistry Research, 49,* 5132–5140.

Huilin, L. (2017). *Kinetic theory of dense fluid-particle two-phase flows.* Beijing, China (in Chinese): Science Press.

Khongprom, P., & Gidaspow, D. (2010). Compact fluidized bed sorber for $CO_2$ capture. *Particuology, 8,* 531–535.

Rao, A., Curtis, J. S., Hancock, B. C., & Wassgren, C. (2011). Classifying the fluidization and segregation behavior of binary mixtures using particle size and density ratios. *AIChE Journal, 57,* 1446–1458.

Shuai, W., Guodong, L., Huilin, L., Juhui, C., Yurong, H., & Jiaxing, W. (2011a). Fluid dynamic simulation in a chemical looping combustion with two interconnected fluidized beds. *Fuel Processing Technology, 92,* 385–393.

Shuai, W., Yunchao, Y., Huilin, L., Pengfei, X., & Liyan, S. (2012). Computational fluid dynamic simulation based cluster structures-dependent drag coefficient model in dual circulating fluidized beds of chemical looping combustion. *Industrial and Engineering Chemistry Research, 51,* 1396–1412.

Shuai, W., Huilin, L., Feixiang, Z., & Guodong, L. (2014a). CFD studies of dual circulating fluidized bed reactors for chemical looping combustion processes. *Chemical Engineering Journal, 236,* 121–130.

Shuai, W., Huang, L., Zhenhua, H., Huilin, L., Guodong, L., Dan, L., & Feixiang, Z. (2014b). Numerical modeling of a bubbling fluidized bed coal gasifier by kinetic theory of rough spheres. *Fuel, 130,* 197–202.

Shuai, W., Liming, Y., Feixiang, Z., Huilin, L., Liyan, S., & Qinghong, Z. (2014c). Numerical simulation of hydrogen production via chemical looping reforming in interconnected fluidized bed reactor. *Industrial and Engineering Chemistry Research, 53,* 4182–4191.

Shuai, W., Juhui, C., Huilin, L., Guodong, L., & Liyan, S. (2015a). Multi-scale simulation of chemical looping combustion in dual circulating fluidized bed. *Applied Energy, 155,* 719–727.

Shuai, W., Qi, W., Juhui, C., Guodong, L., Huilin, L., & Liyan, S. (2016a). Assessment of $CO_2$ capture using potassium-based sorbents in circulating fluidized bed reactor by multiscale modeling. *Fuel, 164,* 66–72.

Shuai, W., Xiaojiao, S., Juhui, C., Qi, W., & Huilin, L. (2016b). Insights in steam reforming of glycerol in a fluidized bed by computational fluid dynamics modeling. *Energy and Fuels, 30,* 8335–8342.

Shuyan, W., Lijie, Y., Huilin, L., Jianmin, D., Long, Y., & Xiang, L. (2008a). Numerical analysis of particle clustering effects on desulphurization and NO emission in a circulating fluidized bed combustor. *Fuel, 87,* 870–877.

Shuyan, W., Lijie, Y., Huilin, L., Yurong, H., Jianmian, D., Guodong, L., & Xiang, L. (2008b). Simulation of effect of catalytic particle clustering on methane steam reforming in a circulating fluidized bed reformer. *Chemical Engineering Journal, 139,* 136–146.

Wen, C. Y., & Yu, Y. U. (1966). A generalized method for predicting the minimum fluidization velocity. *AIChE Journal, 12,* 610–612.

Wu, S. Y., & Baeyens, J. (1998). Segregation by size difference in gas fluidized beds. *Powder Technology, 98,* 139–150.

Yunhau, Z., Huilin, L., Yurong, H., Jianmin, D., & Lijie, Y. (2006). Numerical prediction of combustion of carbon particle clusters in a circulating fluidized bed riser. *Chemical Engineering Journal, 118,* 1–10.